CAMBRIDGE LIBRARY COLLECTION

Books of enduring scholarly value

Monographs of the Palaeontographical Society

The Palaeontographical Society was established in 1847, and is the oldest society devoted to study of palaeontology worldwide. Its primary role is to promote the description and illustration of the British fossil flora and fauna, via publication of an authoritative monograph series. These monographs cover a wide range of taxonomic groups, from microfossils, trilobites and ammonites through to Coal Measure plants, mammals and reptiles, and from all ages from Cambrian to Pleistocene. They form a benchmark for understanding the past life of the British Isles and many include the original descriptions of numerous key species. The first monograph (on the Crag Mollusca) was published in March 1848 and the Society still continues this work today. Notable authors in the series include Charles Darwin (fossil barnacles) and Richard Owen (dinosaurs and other extinct reptiles). Beginning in 2014, the Cambridge Library Collection and the Society are collaborating to reissue the earlier publications, focusing on monographs completed between 1848 and 1918.

The Fossil Fishes of the English Wealden and Purbeck Formations

The Purbeck and Wealden formations of southern England represent marginal marine and continental deposition during the latest Jurassic and Early Cretaceous. More famous for their fossil dinosaurs and mammals, these units also yield the remains of fishes. In this work, first published in three parts between 1916 and 1919, Arthur Smith Woodward (1864–1944) provides the most extensive overview of the Purbeck and Wealden ichthyofauna, describing and illustrating some thirty genera of cartilaginous, lobe-finned, and ray-finned fishes. Woodward finds the preservation of fishes from both deposits to be suboptimal, but nevertheless comes to some important conclusions: he shows that the fish fauna of the English Wealden is nearly identical to that of the famous coeval deposits of Bernissart in Belgium, and finds that the species from both the Wealden and Purbeck show closer affinities with Jurassic forms than with later Cretaceous lineages like those described in his monograph on fishes from the Chalk.

Cambridge University Press has long been a pioneer in the reissuing of out-of-print titles from its own backlist, producing digital reprints of books that are still sought after by scholars and students but could not be reprinted economically using traditional technology. The Cambridge Library Collection extends this activity to a wider range of books which are still of importance to researchers and professionals, either for the source material they contain, or as landmarks in the history of their academic discipline.

Drawing from the world-renowned collections in the Cambridge University Library and other partner libraries, and guided by the advice of experts in each subject area, Cambridge University Press is using state-of-the-art scanning machines in its own Printing House to capture the content of each book selected for inclusion. The files are processed to give a consistently clear, crisp image, and the books finished to the high quality standard for which the Press is recognised around the world. The latest print-on-demand technology ensures that the books will remain available indefinitely, and that orders for single or multiple copies can quickly be supplied.

The Cambridge Library Collection brings back to life books of enduring scholarly value (including out-of-copyright works originally issued by other publishers) across a wide range of disciplines in the humanities and social sciences and in science and technology.

The Fossil Fishes
of the English Wealden and Purbeck Formations

Arthur Smith Woodward

CAMBRIDGE
UNIVERSITY PRESS

University Printing House, Cambridge, CB2 8BS, United Kingdom

Cambridge University Press is part of the University of Cambridge.
It furthers the University's mission by disseminating knowledge in the pursuit of
education, learning and research at the highest international levels of excellence.

www.cambridge.org
Information on this title: www.cambridge.org/9781108076944

© in this compilation Cambridge University Press 2014

This edition first published 1916–19
This digitally printed version 2014

ISBN 978-1-108-07694-4 Paperback

This book reproduces the text of the original edition. The content and language reflect
the beliefs, practices and terminology of their time, and have not been updated.

Cambridge University Press wishes to make clear that the book, unless originally published
by Cambridge, is not being republished by, in association or collaboration with,
or with the endorsement or approval of, the original publisher or its successors in title.

THE

PALÆONTOGRAPHICAL SOCIETY.

INSTITUTED MDCCCXLVII.

VOLUME FOR 1917.

LONDON:

MDCCCCXIX.

THE FOSSIL FISHES OF THE ENGLISH WEALDEN AND PURBECK FORMATIONS.

ORDER OF BINDING AND DATES OF PUBLICATION.

PAGES	PLATES	ISSUED IN VOL. FOR YEAR	PUBLISHED
Title-page and Index	—	1917	April, 1919
1—48	I—X	1915	October, 1916
49—104	XI—XX	1916	February, 1918
105—148	XXI—XXVI	1917	April, 1919

THE
FOSSIL FISHES

OF THE

ENGLISH

WEALDEN AND PURBECK FORMATIONS.

BY

ARTHUR SMITH WOODWARD, LL.D., F.R.S.,

KEEPER OF THE DEPARTMENT OF GEOLOGY IN THE BRITISH MUSEUM; SECRETARY OF THE
PALÆONTOGRAPHICAL SOCIETY.

LONDON:
PRINTED FOR THE PALÆONTOGRAPHICAL SOCIETY.
1916—1919.

PRINTED BY ADLARD AND SON AND WEST NEWMAN, LTD., LONDON AND DORKING.

SYSTEMATIC INDEX.

	PAGE
INTRODUCTION	1
BIBLIOGRAPHY	2
SYSTEMATIC DESCRIPTIONS	3
Subclass ELASMOBRANCHII	3
Order SELACHII	3
Family CESTRACIONTIDÆ	3
Hybodus	3
— basanus	5
— ensis	11
— parvidens	12
— striatulus	12
— strictus	13
Hybodont Cephalic Spines	13
Acrodus	14
— ornatus	14
Asteracanthus	16
— verrucosus	16
— semiverrucosus	17
— granulosus	18
Hylæobatis	19
— problematica	19
Subclass TELEOSTOMI	21
Order CROSSOPTERYGII	21
Family CŒLACANTHIDÆ	21
Undina	21
— purbeckensis	22
Order ACTINOPTERYGII	23
Family PALÆONISCIDÆ	23
Coccolepis	23
— andrewsi	24
— sp.	25
Family SEMIONOTIDÆ	26
Lepidotus	26
— minor	27
— notopterus	34
— mantelli	36
Family PYCNODONTIDÆ	47
Athrodon	47

	PAGE
Subclass TELEOSTOMI (cont.)	
Order ACTINOPTERYGII (cont.)	
Family PYCNODONTIDÆ (cont.)	
Athrodon intermedius	48
Mesodon	48
— daviesi	50
— parvus	52
Eomesodon	54
— barnesi	56
— depressus	57
Microdon	58
— radiatus	59
Cœlodus	64
— mantelli	66
— multidens	67
— hirudo	68
— lævidens	69
— arcuatus	70
Family MACROSEMIIDÆ	70
Ophiopsis	70
— penicillata	71
— breviceps	73
— dorsalis	75
Histionotus	76
— angularis	77
Enchelyolepis	80
— andrewsi	80
Family EUGNATHIDÆ	82
Caturus	82
— (Callopterus ?) latidens	83
— purbeckensis	85
— tenuidens	86
Neorhombolepis	87
— valdensis	87
Family AMIIDÆ	90
Amiopsis	90
— damoni	91

SYSTEMATIC INDEX.

	PAGE
Subclass TELEOSTOMI (*cont.*)	
Order ACTINOPTERYGII (*cont.*)	
Family AMIIDÆ (*cont.*)	
Amiopsis austeni	94
Family ASPIDORHYNCHIDÆ	95
Aspidorhynchus	95
— fisheri	97
Belonostomus	100
— hooleyi	100
Family PHOLIDOPHORIDÆ	101
Pholidophorus	101
— ornatus	102
— granulatus	106
— purbeckensis	108
— brevis	110
Ceramurus	111
— macrocephalus	111
Pleuropholis	113
— attenuata	114
— formosa	115
— crassicauda	118
— longicauda	119
— serrata	120

	PAGE
Subclass TELEOSTOMI (*cont.*)	
Order ACTINOPTERYGII (*cont.*)	
Family OLIGOPLEURIDÆ	121
Œonoscopus sp.	121
Family LEPTOLEPIDÆ	121
Leptolepis	121
— brodiei	122
Æthalion	125
— valdensis	126
Pachythrissops	128
— lævis	129
— vectensis	133
Thrissops	136
— curtus	137
— molossus	138
SUPPLEMENT	139
Hybodus basanus	139
— sulcatus	139
— subcarinatus	140
Lepidotus mantelli	140
— minor	140
SUMMARY AND CONCLUSION	141

LIST OF TEXT-FIGURES.

FIG.		PAGE
1.	*Hybodus hauffianus*; fish showing soft parts	4
2.	Ditto; restoration of skeleton	5
3.	*Hybodus basanus*; restoration of skull	7
4.	Ditto; teeth	8
5.	Ditto; fragment of trunk	9
6.	*Acrodus anningiæ*; dentition	15
7.	*Asteracanthus verrucosus*; dorsal fin-spine	16
8.	*Asteracanthus semiverrucosus*; dorsal fin-spine	17
9.	*Asteracanthus granulosus*; dorsal fin-spine	18
10.	*Hylæobatis problematica*; section of tooth, magnified	20
11.	*Ptychodus mammillaris*; section of tooth, magnified	21
12.	*Lepidotus semiserratus*; restoration of head	27
13.	*Lepidotus mantelli*; restoration of head	27
14.	*Lepidotus minor*; restoration	28
15.	*Lepidotus* sp.; occipital portion of skull	38
16.	*Lepidotus mantelli*; transverse section of skull	39
17.	Ditto; dentition	41
18.	Ditto; vertebræ	44
19.	*Athrodon intermedius*; splenial dentition	48
20.	*Mesodon macropterus*; restoration	49
21.	*Eomesodon liassicus*; drawing of type specimen	55
22.	*Eomesodon barnesi*; drawing of type specimen	56
23.	*Microdon radiatus*; restoration	60
24.	*Cœlodus costæ*; restoration	65
25.	*Ophiopsis procera*; restoration	71
26.	*Caturus furcatus*; restoration	82
27.	*Callopterus insignis*; restoration	83
28.	*Caturus latidens*; imperfect head	84
29.	*Amiopsis dolloi*; restoration, with scales	90
30.	Ditto; restoration, without scales	91
31.	*Amiopsis austeni*; drawing of type specimen	93
32.	*Aspidorhynchus acutirostris*; restoration	95
33.	*Aspidorhynchus* sp.; hinder portion of skull	96
34.	*Pholidophorus ornatus*; restoration	103
35.	*Pleuropholis attenuata, P. longicauda,* and *P. serrata*; drawings of type specimens	114
36.	*Pleuropholis formosa*; restoration	116
37.	*Leptolepis dubius*; restoration	122
38.	*Æthalion robustus*; restoration	126
39.	*Æthalion valdensis*; drawing of type specimen	127
40.	*Pachythrissops vectensis*; head	134
41.	*Lepidotus minor*; amended restoration	140

THE FOSSIL FISHES

OF THE

ENGLISH WEALDEN AND PURBECK FORMATIONS.

INTRODUCTION.

The fishes of the Wealden and Purbeck formations are of special interest as representing the latest of the typical Jurassic faunas. Certain families and genera range even to Upper Cretaceous horizons, but here they are rare and mingled with a multitude of more modern fishes. The great estuary in which the Wealden and Purbeck beds were deposited must have opened into a sea in which there were none but Jurassic forms; and the only noteworthy features of the fishes discovered in these formations are certain marks of senility and an occasional dwarfing of the species. The remains are usually fragmentary, though most of the ganoids are now known by nearly complete specimens, and there are many pieces showing important osteological characters. The fragments in the Wealden are often much waterworn and abraded, while the better preserved fishes in the Purbeck limestones have frequently been so much crushed that the details of their structure are obscured.

Fish-remains seem to have been first noticed in the Wealden formations by Dr. Gideon A. Mantell, who described them in his early works.[1] A fine collection was also made on the Sussex coast by Mr. Samuel H. Beckles, and important series of Wealden specimens have been obtained during more recent years by Messrs. Charles Dawson, Philip Rufford, E. J. Baily, and Reginald W. Hooley. The Mantell, Beckles, Dawson, and Rufford collections are now in the British Museum, while that of Mr. Baily is in the Hastings Museum. Fossil fishes from the Purbeck Beds of Swanage, Dorset, were noticed at least so long ago as 1816[2] and are preserved in many museums. They are especially well represented in the British Museum, the Museum of Practical Geology, and the Dorset County

[1] G. A. Mantell, 'The Fossils of the South Downs' (1822), pp. 45, 46 (description only); 'Illustrations of the Geology of Sussex' (1827).

[2] T. Webster in H. C. Englefield, 'The Isle of Wight' (London, 1816), p. 192.

Museum at Dorchester. Similar fishes from the Purbeck Beds of the Vale of Wardour, Wiltshire, were collected many years ago by the Rev. P. B. Brodie, and more recently by the Rev. W. R. Andrews and Mr. T. T. Gething. All their finest specimens are now in the British Museum and the Museum of Practical Geology. There are also a few fish-remains from the Purbeck Beds of Buckinghamshire in the John Lee Collection at Hartwell House, near Aylesbury.

BIBLIOGRAPHY.

1. AGASSIZ, L.—'Recherches sur les Poissons Fossiles,' vols. i—v. Neuchâtel, 1833—44.
2. BRANCO, W.—"Beiträge zur Kenntniss der Gattung *Lepidotus*," 'Abhandl. k. preuss. geol. Landes-Anstalt,' vol. vii, pt. 4 (1887).
3. BRODIE, P. B.—'A History of the Fossil Insects in the Secondary Rocks of England.' London, 1845. [With notes on fossil fishes by Egerton.]
4. DAVIES, W.—"A New Species of *Pholidophorus* from the Purbeck Beds of Dorsetshire," 'Geol. Mag.' [3], vol. iv. (1887), pp. 337—339, pl. x.
5. EGERTON, P. M. G.—"Description of the Mouth of a *Hybodus* found by Mr. Boscawen Ibbetson in the Isle of Wight," 'Quart. Journ. Geol. Soc.,' vol. i. (1845), pp. 197—199, pl. iv.
6. —— "On some new Genera and Species of Fossil Fishes," 'Ann. Mag. Nat. Hist.' [2], vol. xiii (1854), pp. 433—436. [Abstract of following work.]
7. —— 'Memoirs of the Geological Survey of the United Kingdom: Figures and Descriptions Illustrative of British Organic Remains,' dec. viii (1855).
8. MANSEL-PLEYDELL, J. C.—"On a New Specimen of *Histionotus angularis*, Egerton," 'Geol. Mag.' [3], vol. vi (1889), pp. 241, 242, pl. vii.
9. MANTELL, G. A.—"Illustrations of the Geology of Sussex, with Figures and Descriptions of the Fossils of Tilgate Forest.' London, 1827.
10. REID, C., and STRAHAN, A.—"The Geology of the Isle of Wight," Second Edition, 'Mem. Geol. Surv.,' 1899.
11. STRAHAN, A.—"The Geology of the Isle of Purbeck and Weymouth," 'Mem. Geol. Surv.,' 1898.
12. TOPLEY, W.—"The Geology of the Weald," 'Mem. Geol. Surv.,' 1875.
13. WOODWARD, A. S.—'Catalogue of Fossil Fishes in the British Museum,' Parts I—III (1889—1895).
14. —— "On some New Fishes from the English Wealden and Purbeck Beds, referable to the Genera *Oligopleurus*, *Strobilodus*, and *Mesodon*," 'Proc. Zool. Soc.,' 1890, pp. 346—353, pls. xxviii, xxix.

15. WOODWARD, A. S.—" On the Cranial Osteology of the Mesozoic Ganoid Fishes, *Lepidotus* and *Dapedius*," 'Proc. Zool. Soc.,' 1893, pp. 559—565, pls. xlix, l.
16. —— " A Contribution to Knowledge of the Fossil Fish Fauna of the English Purbeck Beds," 'Geol. Mag.' [4], vol. ii (1895), pp. 145—152, pl. vii.
17. —— " A Description of *Ceramurus macrocephalus*," 'Geol. Mag.' [4], vol. ii (1895), pp. 401—402.
18. —— " Note on the Affinities of the English Wealden Fish-Fauna," 'Geol. Mag.' [4], vol. iii (1896), pp. 69—71.
19. —— " On a new Leptolepid Fish from the Weald Clay of Southwater, Sussex," 'Ann. Mag. Nat. Hist.' [7], vol. xx (1907), pp. 93—95, pl. i.

The following is the most important work on Wealden Fishes from the European Continent.

20. TRAQUAIR, R. H.—" Les Poissons Wealdiens de Bernissart," 'Mém. Mus. roy. d'Hist. nat. Belgique,' vol. vi (1911).

SYSTEMATIC DESCRIPTIONS.

Subclass *ELASMOBRANCHII*.

Order *SELACHII*.

Family CESTRACIONTIDÆ.

Genus **HYBODUS**, Agassiz.

Hybodus, L. Agassiz, Poiss. Foss., vol. iii, 1837, p. 41.
Sphenonchus, L. Agassiz, op. cit., vol. iii, 1843, p. 201 (in part).
Meristodon, L. Agassiz, op. cit., vol. iii, 1843, p. 286.

Generic Characters.—Trunk fusiform, moderately elongated; the first dorsal fin opposite to the space between the pectoral and pelvic fins, the second in advance of the anal fin. Snout not prominent but mouth inferior; pterygo-quadrate cartilage not articulated with the preorbital region of the skull. Teeth conical or cuspidate, the crown more or less striated, with one principal elevation, and one or more lateral prominences on either side diminishing from the centre; root depressed, but not expanded inwards. Symphysial teeth few and large. Notochord persistent; slender ribs, not reaching the ventral border; intercalary cartilages almost or completely absent. Dorsal fin-spines longitudinally ridged and grooved, the ridges not denticulated; posterior denticles in two longitudinal series, often alternating, not marginal but placed close together on a mesial ridge. Shagreen consisting of small conical, radiately-grooved tubercles, sometimes two or three fused together. One or two large hook-shaped dermal spines, each on a triradiate base, immediately behind the orbit, at least in males.

Type Species.—The generic name *Hybodus* appears to have been given by Agassiz first to some teeth from the German Muschelkalk known as *Hybodus plicatilis* (quoted, without description, by F. A. v. Alberti, Jahrb. f. Min., Geogn., etc., 1832, p. 227). It was not defined until he had examined specimens from the Lower Lias of Lyme Regis, Dorsetshire, showing the teeth and dorsal fin-spines in natural association (L. Agassiz, Poiss. Foss., vol. iii, pp. 41, 178). *Hybodus reticulatus*, from that formation and locality, was the first species satisfactorily described, and may therefore be regarded as the type (L. Agassiz, *tom. cit.*, p. 180, pl. xxiv, fig. 26; pl. xxii*a*, figs. 22, 23).

Remarks.—The known specimens of the several species of *Hybodus* from the Lower Lias of Lyme Regis exhibit not only the arrangement of the dentition and

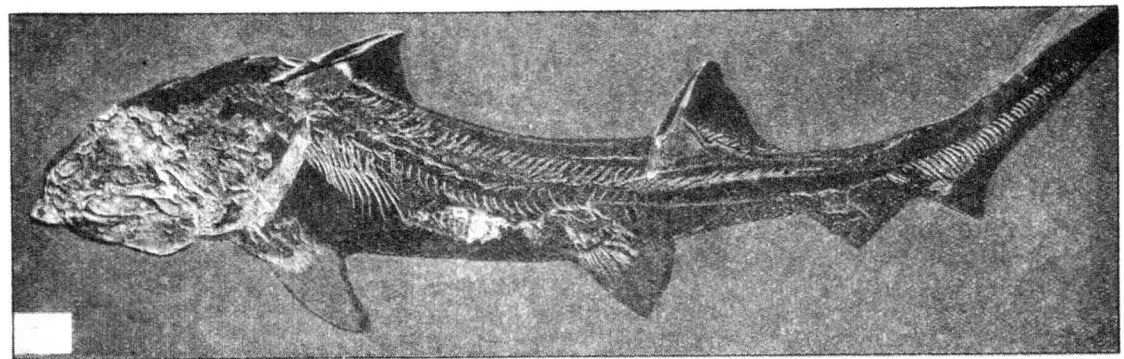

FIG. 1.—*Hybodus hauffianus*, Fraas; fish in left side view, with traces of soft parts, including the fins, about one-fifteenth nat. size.—Upper Lias; Holzmaden, Würtemberg. University Geological Museum, Tübingen.

the dermal armature, but also the cartilages of the jaws, the neural and hæmal arches of the trunk bounding a vacant space for the notochord, and the cartilages of the pectoral arch. Specimens of another species from the Upper Lias of Würtemberg are still more satisfactory, and one example prepared by Mr. Bernhard Hauff shows distinct remains even of the fins (Text-fig. 1).[1] A specimen from the Lithographic Stone (Lower Kimmeridgian) of Bavaria displays the five branchial arches and the cartilages of the pectoral fin.[2] A more imperfect specimen from the same formation in the Montsech, Lérida, Spain, shows the neural arches, slender ribs, and the cartilaginous support of the anterior dorsal fin.[3] Another fragment from the Upper Beaufort Beds of Orange River Colony, South Africa, exhibits the supports of a dorsal fin.[4] The well-preserved skulls and portions of

[1] *Hybodus hauffianus*, E. Fraas, E. Koken, Geol. u. Palæont. Abhandl., n. s., vol. v (1907), pp. 261—276, pls. xi—xiii.

[2] *Hybodus fraasi*, C. Brown, Palæontographica, vol. xlvi (1900), pp. 151—158, pl. xv.

[3] *Hybodus woodwardi*, L. M. Vidal, Bol. Inst. Geol. España, 1915, p. 22, pl. ii, text-figs. 4—6.

[4] *Hybodus africanus*, R. Broom, Ann. S. African Museum, vol. vii (1909), p. 252, pl. xii, fig. 2.

trunk of *Hybodus basanus* from the Wealden of the Isle of Wight and Sussex show still better the shape of the jaws and branchial arches, besides the usual notochordal axial skeleton of the trunk, and the supports of the two dorsal fins.

Hybodus basanus is the only Wealden species sufficiently well known for definition. The other Wealden and Purbeck species are represented by isolated teeth and spines, which bear merely provisional names.

1. **Hybodus basanus,** Egerton. Plate I, figs. 1, 2; Plate II, fig. 1; Text-figures 3—5.

1845. *Hybodus basanus*, P. M. G. Egerton, Quart. Journ. Geol. Soc., vol. i, p. 197, pl. iv.
1889. *Hybodus basanus*, A. S. Woodward, Catal. Foss. Fishes B. M., pt. i, p. 273, pl. xii, figs. 1–5.
1891. *Hybodus basanus*, A. S. Woodward, Proc. Yorks. Geol. Polyt. Soc., vol. xii, p. 63, pl. i; pl. ii, fig. 1.

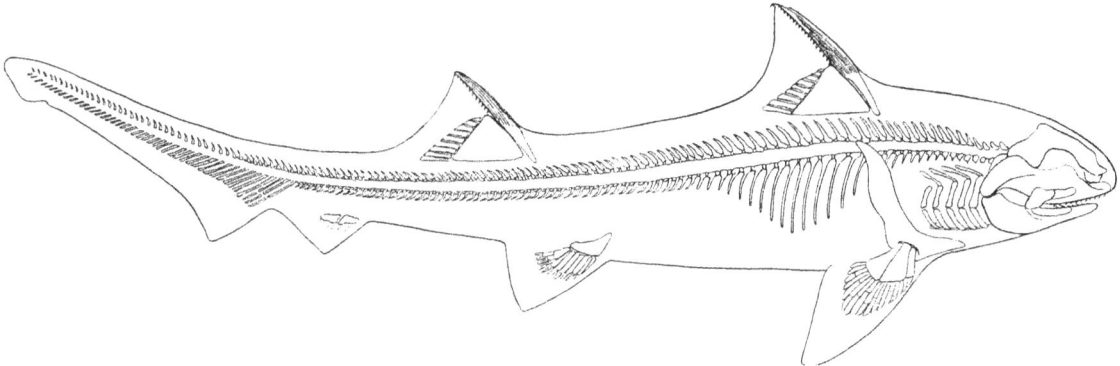

Fig. 2.—*Hybodus hauffianus*, Fraas; restoration of skeleton, about one-fifteenth nat. size.—Upper Lias; Holzmaden, Würtemberg.

Type.—Imperfect skull and mandible with dentition; Museum of Practical Geology, Jermyn Street, London.

Specific Characters.—Teeth with a very high, much compressed crown; median cone, narrow, slender, slightly arched inwards; lateral cones two, sometimes with a rudiment of a third, short but sharply pointed; coronal surface marked by numerous very fine vertical wrinkles, often extending to the apices of the lateral cones, but always absent on the smooth upper portion of the median cone. Dorsal fin-spines rather slender and not much arched, laterally compressed, with a sharp anterior keel; lateral face of exserted portion completely covered with sharp but fine longitudinal ridges, about eight being widely spaced, and those near the posterior border closely arranged; inserted base slender and tapering, its anterior border sometimes inclined at an angle to that of the exserted portion. A single pair of large postorbital cephalic spines with a terminal barb. Conical dermal granules small and fluted.

Description of Specimens.—The type specimen is an imperfect skull and mandible discovered by Capt. L. L. Boscawen Ibbetson at the top of the Wealden near Atherfield, Isle of Wight, and is now in the Museum of Practical Geology. Since its description by Egerton (*loc. cit.*, 1845), it has been cleaned from the matrix, and new drawings of the specimen from the right side and from below are given in Pl. I, figs. 1, 1 *a*. All the cartilages are distorted by crushing, and the waterworn teeth are less distinct than indicated in Egerton's original figure, where each principal cusp appears too wide and smooth.

All the other known specimens are similar heads and fragments of the trunk picked up on the beach of Pevensey Bay, Sussex, where a large collection, now in the British Museum, was made by Mr. S. H. Beckles. From these fossils the principal characters of the species and several interesting anatomical features can be determined.

The cartilages agree with those of modern sharks in being only superficially calcified in the usual small polygonal tesseræ. They are therefore often distorted, not merely by crushing during fossilisation, but also by contraction before burial in the sediment. Under such circumstances their state of preservation is remarkable. In most cases the hollow left by the decay of the internal uncalcified cartilage is filled with ordinary matrix; but sometimes (as in the original of Pl. II, fig. 1) it still remains partly vacant.

The cranium as shown in the type specimen (Pl. I, fig. 1) is rather short and wide, with a relatively large orbit (*orb.*), short postorbital and rostral regions, and a large anterior fontanelle (*a.f*). Its special features, however, are better seen in other specimens, particularly in the unique skull represented in Pl. II, figs. 1, 1 *a*, 1 *b*. This lacks only the occipital region, which is preserved in another specimen in the Enniskillen Collection (B. M. no. P. 3172 *c*), and is seen to slope backwards and downwards, while it is raised in the middle into a sharp vertical ridge extending from the occipital border to the foramen magnum. The cranial roof throughout its length (Pl. II, fig. 1) is gently convex from side to side, is produced downwards into a large postorbital process (fig. 1 *a*), and extends above the orbit into a thin supraorbital flange, which merges in front into the depressed and only slightly expanded region of the nasal capsule. In the middle of the roof of the postorbital region the posterior fontanelle (*p.f.*) is elongate-oval in shape. In front of and between the nasal capsules, the large anterior fontanelle (*a.f.*) is much broader than deep and is directed forwards; while the flat base of the mesethmoidal region soon terminates in a very short but well-marked rostral prominence (*r.*).

The mandibular suspensorium is inclined backwards, so that, since the jaws extend forwards as far as the end of the snout, they are longer than the cranium. As shown by the type specimen (Pl. I, fig. 1) they are also relatively large and massive, with labial cartilages at the angle of the mouth. The hyomandibular (Pl. II, figs. 1, 1 *a*, *hm.*) is a comparatively slender cartilage, laterally compressed and produced somewhat forwards at its upper end, antero-posteriorly compressed but less expanded at its lower end. The pterygo-quadrate (as seen especially in Pl. II, figs. 1, 1 a (*ptq.*), and in the specimen figured in Catal. Foss. Fishes, Brit. Mus., pt. i, pl. xii, fig. 1) is weak and depressed at its anterior symphysis, but deepens rapidly backwards, so that by the middle of the orbit its depth equals at least a quarter of its total length. Its upper border is then slightly concave, and finally rises a little to its highest point behind. It can scarcely have articulated with the postorbital prominence of the cranium. The outer face of its posterior half is indented below, and this hollow is overhung by an arched ridge which runs upwards and forwards from the articular end and dies out before reaching the

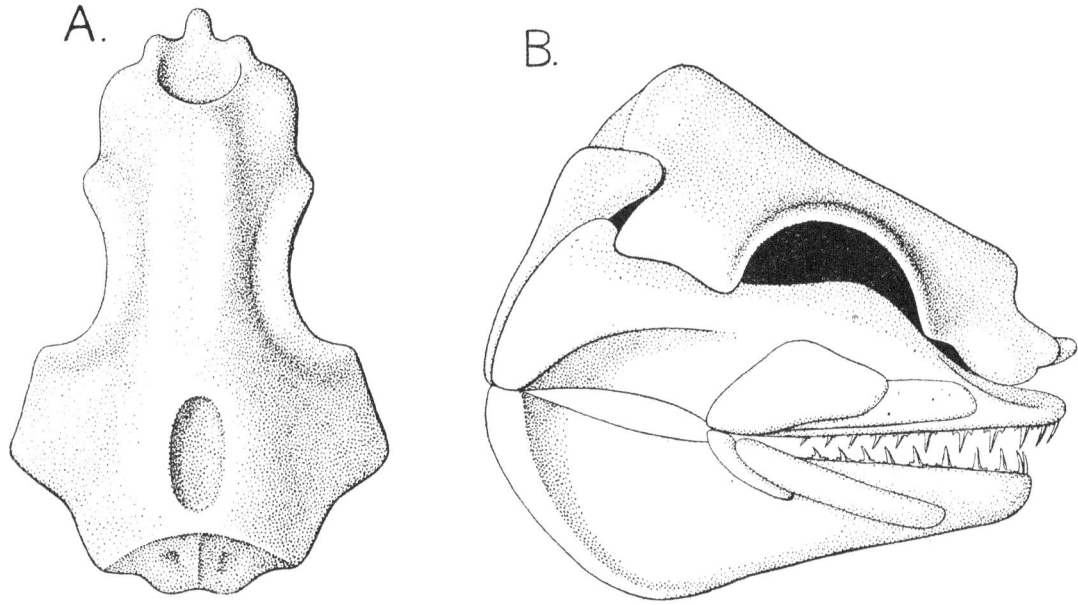

Fig. 3.—*Hybodus basanus*, Egerton; restoration of cranium, upper view (A), and of skull with jaws, right side view (B), about one-half nat. size.—Weald Clay; Pevensey Bay, Sussex.

upper border. The rami of the mandible (Pl. I, figs. 1, 1 a; Pl. II, figs. 1 a, 1 b; *md.*), though deep and massive behind, rapidly taper forwards and meet in a comparatively feeble symphysis, which does not extend so far as the front of the upper jaw. There are two pairs of large labial cartilages, best shown in Pl. II, figs. 1 a, 1 b (*u.l.* 1, 2, *l.l.* 1, 2). Those of the upper and lower anterior pairs are long and band-like; those of the upper posterior pair (Pl. II, fig. 1 a, *u.l.* 2) are large, irregular laminæ; while those of the lower posterior pair (Pl. II, fig. 1 b, *l.l.* 2) are short but stout rounded rods. An attempted restoration of the skull with jaws is given in Text-fig. 3.

As shown by the type specimen (Pl. I, fig. 1) the teeth are in contact round the margin of the jaws, and at least three or four series, one behind the other, must have been simultaneously in use. An examination of several specimens proves that each ramus in both jaws bears ten or eleven transverse rows of teeth; while one skull in the British Museum (no. P. 3172 a) seems to exhibit an unpaired symphysial row in the lower jaw. In all the teeth the principal cusp is high and narrow, compressed to two sharp lateral edges, with the incurved apex smooth and the expanded base vertically striated. The one or two pairs of well-defined lateral denticles are striated to the apex. From the symphysis backwards to the middle of each ramus the teeth are highest and about equally elevated; but in this series those at and near the symphysis have a less extended base than those further back, with less space for the lateral denticles, which are usually in two pairs (the outer very small), but may be flanked by a third minute cusp (Pl. I, fig. 1 b). In the hinder half of each ramus the teeth rapidly diminish in size and elevation, with the principal cusp curving sharply backwards (Text-fig. 4). There is no essential difference between the teeth of the upper and lower jaws.

The ceratohyals are massive cartilages seen in several specimens, and the

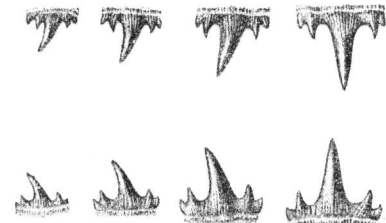

Fig. 4.—*Hybodus basanus*, Egerton; four upper and lower teeth from hinder half of jaws, nat. size.—Weald Clay; Pevensey Bay, Sussex.

basihyal is also large, somewhat broader than long, as already described in Catal. Foss. Fishes, Brit. Mus., pt. i (1889), p. 274, pl. xii, fig. 2. The branchial arches are only five in number, as shown by the ceratobranchials preserved in series in a specimen already described, *loc. cit.*, p. 274, pl. xii, fig. 3, and as still better seen in another head in the Beckles Collection (B. M. no. P. 11872). The hindmost or fifth arch is comparatively small. Each ceratobranchial is expanded and sharply truncated at its lower end, where it would articulate with the hypobranchial; but the cartilages of this lower series remain undiscovered.

The trunk is known only by fragments, of which the best is represented in Text-fig. 5. The notochord must have been persistent, but the neural arches and spines (*n.s.*) are well calcified in the usual granular form. They are narrow bands of cartilage arranged in close series. Below the space for the notochord in the abdominal region there are also traces of comparatively slender hæmal elements or ribs in a specimen described in Proc. Yorks. Geol. Polyt. Soc., vol. xii (1891), p. 65, pl. ii, fig. 1.

Of the fins, only parts of the dorsals have hitherto been discovered in the original of Text-fig. 5. In this specimen the anterior dorsal (*d*. 1) probably remains in its natural position, but the second dorsal (*d*. 2) is accidentally overturned and displaced. Each is shown to have been supported in the usual manner by a dorsal fin-spine fixed to a triangular basal cartilage, which extends from the inserted end of the spine throughout the whole length of the fissure on its hinder face. At the distal border of the basal cartilage of the posterior fin five small radials also occur, gradually increasing in length towards the hinder edge of the fin; and there are traces of delicate filiform rays for the support of the fin-membrane. As in *Hybodus fraasi* and *H. hauffianus*, the basal cartilage of the anterior fin is narrower and deeper than that of the posterior fin.

The dorsal fin-spines are much laterally compressed and very little arched, with a comparatively slender base of insertion. The sides of the exserted portion

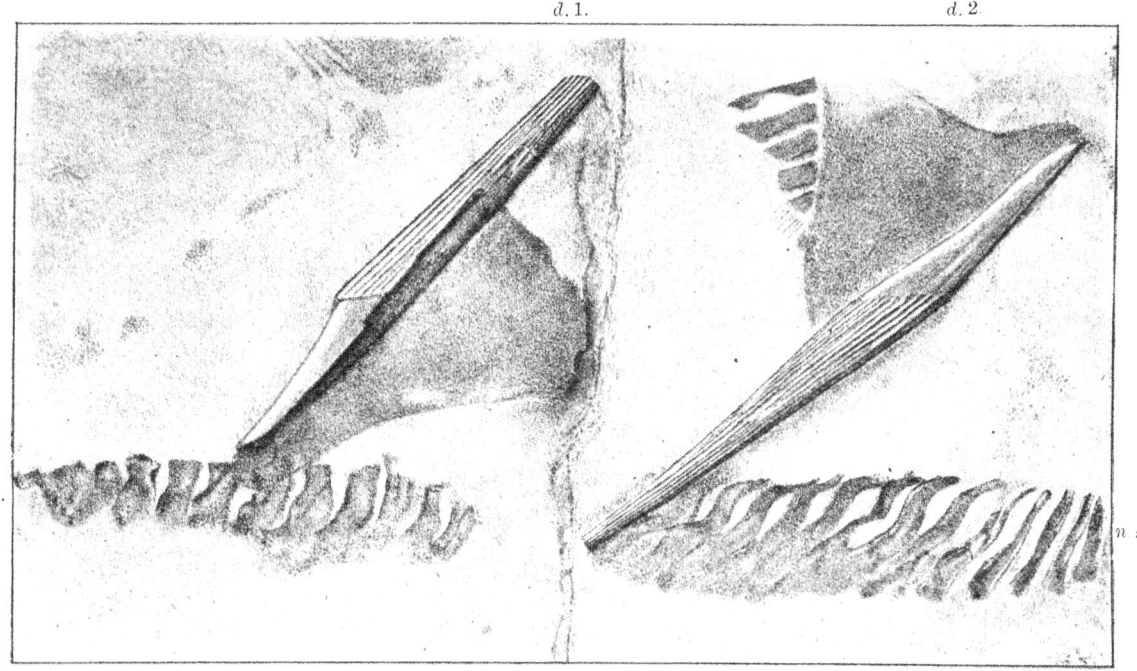

Fig. 5.—*Hybodus basanus*, Egerton; fragment of trunk in left side view, showing the neural arches (*n.s.*) of the vertebral axis, with the spines and cartilages of the anterior (*d*. 1) and posterior (*d*. 2) dorsal fins, the latter overturned and displaced, nearly one-half nat. size.—Weald Clay: Pevensey Bay, Sussex. Beckles Collection (B. M. no. P. 6357).

are completely covered with fine and sharp longitudinal ridges, which are sometimes slightly nodulose where crossed by growth-lines. Near the base about eight ridges are widely spaced, while four or five at the posterior border are crowded. The posterior denticles are numerous, small, and closely arranged. As shown in Text-fig. 5, the spine of the anterior fin is broader than that of the posterior fin.

In some specimens, which are probably to be regarded as males, there is also a single pair of spines immediately behind the head. This is best shown in the partially decayed skull represented in Pl. I, fig. 2. The spine (*s.*) is placed laterally just behind the position of the hyomandibular, and seems to have been fixed on a special cartilaginous support (*x.*). It is of the form originally named *Sphenonchus*, with a trifid inserted base, from which rises a sigmoidally arched enamelled spine, barbed at the apex (fig. 2 *a*). As observed in B. M. no. P. 11872, the postero-inferior limb of the base is largest and longest and truncated at the end, while its long axis is slightly oblique to that of the other two limbs, which are nearly in the same line but curved. The exserted spine, which rises as usual at the place of meeting of the three basal limbs, is at least as long as the postero-inferior limb, laterally compressed, and sufficiently unsymmetrical to show that it is not a median structure. It is completely covered with enamel, which is smooth at the barbed apex and along the narrow upper face; but its basal portion is marked with irregular sharp ridges, which cover the greater part of the antero-lateral face and here terminate abruptly above at a sharp longitudinal ridge which extends to the apex.

The head and at least the anterior portion of the trunk are covered with a spinous shagreen, which is always fine, but varies a little in size in different regions. Each tubercle (Pl. II, figs. 1 *c*, 1 *d*) is hollow, with an expanded trumpet-shaped base, more or less crimped round the edge, and marked with radiating ridges on its outer face. It rises in the middle into a laterally-compressed recurved hooklet, on which the vertical ridges end abruptly at the arched anterior border.

Horizon and Localities.—Weald Clay: Atherfield, Isle of Wight; Cooden Beach, Pevensey Bay, Sussex.

Addendum.—Isolated teeth of the same general type as those of *H. basanus* also occur in lower horizons of the Wealden series, but are not sufficiently similar to be referred with certainty to this species. Some obtained by Mantell from the Tunbridge Wells Sand of Tilgate Forest seem to have the principal cone less compressed and the inner lateral denticles more slender and acuminate than in *H. basanus* (as shown in Catal. Foss. Fishes, Brit. Mus., pt. i, 1889, pl. xi, figs. 14, 15). Rolled and waterworn fragments of such teeth were named *Oxyrhina (Meristodon) paradoxa* or *Meristodon paradoxus* by Agassiz, Poiss. Foss., vol. iii (1843), p. 286, pl. xxxvi, figs. 53—56.

Abraded and fragmentary small dorsal fin-spines from the Tunbridge Wells Sand of Tilgate Forest also closely resemble those of *H. basanus*, but can scarcely be described as identical. They were named *Hybodus subcarinatus* by Agassiz, Poiss. Foss., vol. iii (1837), p. 46, pl. x, figs. 10—12; and an early figure of one specimen was given in Trans. Geol. Soc. [2], vol. ii (1829), pl. vi, fig. 9. Two nearly complete fin-spines, less abraded than usual, from the Wadhurst Clay near Hastings, are shown in Pl. III, figs. 6, 7, and appear to be essentially identical with those of *H. basanus*, having the same fine longitudinal ridges and small posterior denticles. The broader spine (fig. 6) is a little widened by crushing; the narrower spine (fig. 7) exhibits only the broken bases of the posterior denticles.

2. Hybodus ensis, sp. nov. Plate II, figs. 2—7.

Type.—Tooth; British Museum.

Specific Characters.—Teeth sometimes 2 cm. in diameter, longest usually smaller. Median cone high, much compressed, broad at the base, tapering gradually to a blunt apex; two or three lateral cones, slender and sharply pointed, close to the median cone; coronal surface marked at the base with numerous delicate vertical wrinkles, which nearly reach the apices of the lateral cones.

Description of Specimens.—This species is definitely known only by isolated teeth, of which the original of Pl. II, fig. 6, may be regarded as the type specimen. Here the median cone is complete, except for slight abrasion of its apex; the characteristic slender inner lateral cone is also well shown; and there is a trace of a minute outer cone on one side. Near the base the fine vertical wrinkles are conspicuous, and they do not extend quite to the apex of the inner lateral cones. The original of fig. 3 is a crushed larger tooth of nearly similar form, with the left lateral cone broken away at the apex. A still larger tooth, much abraded, with imperfect lateral cones, is shown in fig. 5. In the tooth represented in fig. 2 the apex of the median cone is blunted by fracture, while in the original of fig. 4 it is complete. Both these teeth have a minute outer pair of lateral cones. Fig. 7 shows a smaller tooth with the median cone much inclined backwards, evidently referable to the hinder part of the jaw. It has three lateral cones in front. In all the teeth the compression of the median cone causes its lateral borders to be especially thin.

The teeth now described have sometimes been referred to the typically Lower Oolitic species, *Hybodus grossiconus*, Ag., but most of them are of smaller size, and they are readily distinguished by the less lateral expansion of their base-line and the somewhat blunter apex of their median cone.

Dorsal Fin-spines.—It is interesting to notice that in the same horizon as the teeth of *Hybodus ensis* there also occur dorsal fin-spines almost identical with those named *H. dorsalis* (L. Agassiz, Poiss. Foss., vol. iii, 1837, p. 42, pl. x, fig. 1), which are found in Bathonian formations with the teeth of *H. grossiconus*. These spines, of which three are shown in Pl. III, figs. 1—3, may therefore possibly belong to *H. ensis*. They are rather stout, with coarser and rounder ridges than the other Hybodont fin-spines met with in Purbeck and Wealden deposits, and their posterior denticles are relatively large. The crushed specimen represented in fig. 1 is short and wide, with regular smooth ribbing, but only traces of the posterior denticles. The original of fig. 2 is an abraded fragment with very coarse and partly nodulose or wavy ridging. Fig. 3 represents a smaller and more elongated spine with large, irregular, hooked posterior denticles.

Horizons and Localities.—Middle Purbeck Beds: Swanage. Wealden: Tilgate Forest.

3. Hybodus parvidens, sp. nov. Pl. II, figs. 8—14.

1889. *Hybodus*, sp. inc., A. S. Woodward, Catal. Foss. Fishes, B. M., pt. i, p. 276, pl. xi, fig. 16.

Type.—Tooth; British Museum.

Specific Characters.—Teeth small, rarely exceeding a centimetre in longest diameter; median cone stout and large, elevated and acute in the anterior teeth, low and blunter in the lateral and posterior teeth; lateral cones two or three on each side, also low and stout; coronal surface marked by sparse vertical wrinkles, which extend to the apices of the lateral cones, some usually also to the apex of the median cone; occasional small excrescences at the base of the crown.

Description of Specimens.—This species is known only by small, isolated teeth, of which the original of Pl. II, fig. 8, may be regarded as the type specimen. Its median cone is only moderately elevated, flanked with two pairs of very blunt lateral cones, and marked with especially prominent and sparse wrinkles. The original of fig. 9 has a broader and stouter median cone, with three imperfectly separated lateral cones on one side, two on the other, all marked with less sparse wrinkles. This tooth passes into those shown in figs. 10 and 12, which have a still stouter and less elevated median cone, and doubtless belong to the back of the jaw. They are noteworthy for the slight arching of their base-line, and a larger tooth of nearly the same form (figured in Catal. Foss. Fishes, B. M., pt. i, pl. xi, fig. 16) shows some traces of basal excrescences. Fig. 13 represents a larger tooth, with three pairs of lateral denticles, perhaps referable to the middle part of the ramus of the jaw. The originals of figs. 11 and 14, with a more elevated median cone and two pairs of relatively small lateral cones, are evidently anterior teeth, and are remarkable for the small excrescence at the middle of the base of the crown.

Though much smaller, these teeth closely resemble those of the Upper Jurassic *Hybodus obtusus*, Ag., in which the small excrescences at the base of the crown are especially numerous and prominent.

Horizons and Localities.—Wealden (chiefly Wadhurst Clay): Hastings. Weald Clay: Berwick, Sussex.

4. Hybodus striatulus, Agassiz. Plate III, fig. 8.

1827. "Resembling *Silurus*," G. A. Mantell, Illustr. Geol. Sussex, p. 58, pl. x, fig. 4.
1837. *Hybodus striatulus*, L. Agassiz, Poiss. Foss., vol. iii, p. 44, pl. viii *b*, fig. 1.

Type.—Portion of dorsal fin-spine; British Museum.
Specific Characters.—Dorsal fin-spines attaining a length of nearly 25 cm.,

stout and not much arched, with slightly rounded sides and blunt anterior keel; lateral face of exserted portion covered with coarse rounded longitudinal ridges, which are closely arranged and in the distal portion tend to become subdivided into tubercles.

Description of Specimens.—Like most of the fossils from the Wealden of Tilgate, the only two known specimens of this form of dorsal fin-spine are much water-worn and abraded. The large spine figured by Mantell (figure copied by Agassiz) is especially abraded, so that traces of the rounded longitudinal ridges (not shown in the published figure) are observable only in the distal half near the front border. The smoothness of the specimen and the bluntness of the posterior denticles are due entirely to abrasion. The second specimen (Pl. III, fig. 8) is part of the distal half of a spine with the ridged ornament better preserved, and interesting as exhibiting a tendency to the subdivision of the ridges into tubercles.

There is considerable resemblance between this form of spine and that mentioned above (p. 11) in connection with *Hybodus ensis;* but the longitudinal ridges in the latter are very rarely nodulose and still more rarely subdivided.

Horizon and Locality.—Tunbridge Wells Sands: Tilgate Forest.

5. **Hybodus strictus,** Agassiz.   Plate III, figs. 4, 5.

1837. *Hybodus strictus*, L. Agassiz, Poiss. Foss., vol. iii, p. 45, pl. x, figs. 7—9.

Type.—Dorsal fin-spine; Bristol Museum.

Specific Characters.—Dorsal fin-spines attaining a length of about 12 or 13 cm., slender and not much arched, laterally compressed, with a sharp anterior keel; lateral face of exserted portion covered with sharp, strong, longitudinal ridges, well spaced except near the lower part of the posterior border, where they are finer and crowded; posterior denticles moderately large; inserted base slender and tapering.

Description of Specimens.—The two examples of this fin-spine shown in Pl. III, figs. 4, 5, are typical, and its characters are very constant in the numerous known specimens. It is of the same general form as the spines named *H. subcarinatus* and *H. basanus*, but is distinguished by its stronger ribs and larger posterior denticles.

Horizon and Locality.—Middle Purbeck Beds: Swanage.

6. **Hybodont Cephalic Spines.**   Plate I, figs. 3, 4.

The cephalic spine already described in *Hybodus basanus* (p. 10) is closely similar in shape to that of the typical *Hybodus* from the Lower Lias; and several portions of spines found isolated both in Wealden and Purbeck formations agree with these in their striated ornament and terminal barb. Two fragments, obtained by Mantell from the Tunbridge Wells Sands of Tilgate Forest and apparently worn smooth by abrasion, were described under the name of *Sphenonchus elongatus* by Agassiz, Poiss. Foss., vol. iii, 1843, p. 202, pl. xxii *a*, figs. 18, 19. Besides these typical cephalic spines, however, there also occur in Wealden beds comparatively small specimens in which the enamelled exserted portion is reduced to a smooth pointed hook without any barb. One is well seen in side view in Pl. I, fig. 3, and a nearly complete example is shown both in side and outer view in Pl. I, figs. 4, 4 *a*. Here the inserted triradiate base is remarkably large, with its postero-inferior limb small and the two lateral limbs much enlarged and inclined downwards. This, indeed, seems to represent the final degenerate condition of the Hybodont cephalic spine.

Genus **ACRODUS**, Agassiz.

Acrodus, L. Agassiz, Poiss. Foss., vol. iii, 1838, p. 139.
Sphenonchus, L. Agassiz, *op. cit.*, vol. iii, 1843, p. 201 (in part).
Thectodus, H. von Meyer and T. Plieninger, Beitr. Paläont. Würtembergs, 1844, p. 116.

Generic Characters.—Only differing from *Hybodus* in the rounded, non-cuspidate shape of the teeth.

Type Species.—The generic name appears to have been given by Agassiz first to the teeth of *Acrodus gaillardoti* from the German Muschelkalk, in Gaillardot, Ann. Sci. Nat., ser. 2, vol. iii (Zoologie), 1835, p. 49, and in Mougeot, Bull. Soc. Géol. France, vol. vi, 1835, p. 20, but it was not defined until the discovery and description of *Acrodus nobilis* from the Lower Lias of Lyme Regis (Agassiz, Poiss. Foss., vol. iii, 1838, p. 140, pl. xxi). The latter must therefore be regarded as the type species.

Remarks.—*Acrodus* is best known by the remains of *A. nobilis* and *A. anningiæ* from the Lower Lias of Lyme Regis, described in Catal. Foss. Fishes, Brit. Mus., pt. i, 1889, pp. 283–295, pls. xiii, xiv. The arrangement of the dentition is shown in Text-fig. 6.

1. **Acrodus ornatus**, A. S. Woodward. Plate II, figs. 15—18.

1889. *Acrodus ornatus*, A. S. Woodward, Catal. Foss. Fishes, Brit. Mus., pt. i, p. 296, pl. xiii, fig. 10.

Type.—Detached tooth; British Museum.
Specific Characters.—A very small species known only by detached teeth, which do not exceed about 7 mm. in length. The dental coronal contour is low

and gently rounded, marked by a longitudinal median wrinkle; the laterally directed wrinkles are short, unusually stout, and marginal, but few tapering and extending to the middle line.

Description of Specimens.—The type tooth from Brixton, Isle of Wight, is elongate-ovoid in shape, about twice as long as wide, probably belonging to one of the principal lateral rows. A second specimen, obtained by the Rev. William Fox from the same locality, is slightly longer and more attenuated at the extremities. A somewhat larger tooth from Brook, of nearly similar shape, broken at one end and naturally curved at the other tapering end, is shown enlarged five

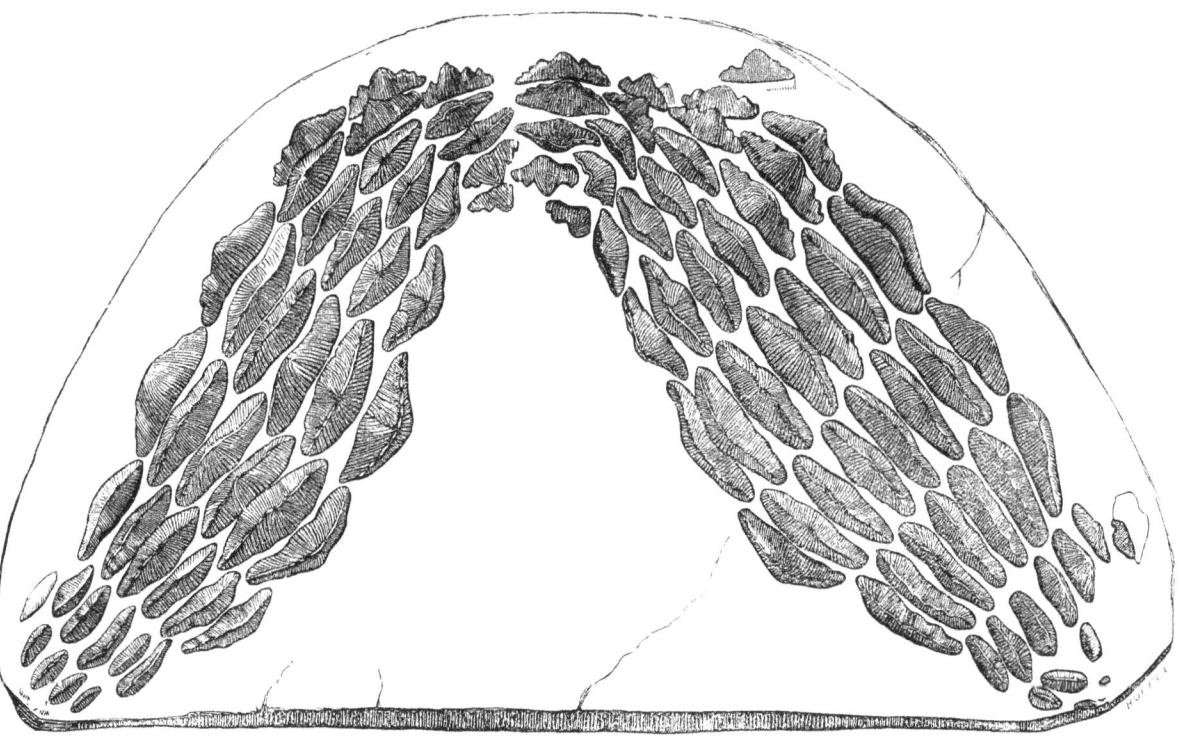

Fig. 6.—*Acrodus anningiæ*, Agassiz; dentition in matrix, almost undisturbed, nat. size.—Lower Lias; Lyme Regis, Dorset. British Museum, no. 39925. After E. C. H. Day, Geol. Mag., vol. i (1864), pl. iii.

times in Pl. II, fig. 15. This is completely unworn, and exhibits well the longitudinal median wrinkle, with the few and thick lateral wrinkles, which taper towards the middle line but rarely reach it. Another specimen of more regular shape (Pl. II, fig. 16), with almost equally well preserved wrinkles, also belongs to a principal row. The relatively short and wide teeth which may be referred to the rows near the symphysis, such as the original of Pl. II, fig. 17, do not exhibit any trace of lateral elevations or denticles. Other teeth, more irregular in shape and wrinkling, probably belong to the hinder part of the jaw (Pl. II, fig. 18).

Horizon and Localities.—Wealden: Brixton and Brook, Isle of Wight; Hastings and Bexhill, Sussex. Waterworn specimens of nearly similar teeth have also been found in the Lower Greensand of Godalming, Surrey.

16 WEALDEN AND PURBECK FOSSIL FISHES.

Genus ASTERACANTHUS, Agassiz.

Asteracanthus, L. Agassiz, Poiss. Foss., vol. iii, 1837, p. 31.
Strophodus, L. Agassiz, op. cit., vol. iii, 1838, p. 116.
Sphenonchus, L. Agassiz, op. cit., vol. iii, 1843, p. 201 (in part).
Curtodus, H. E. Sauvage, Catal. Poiss. Form. Second. Boulonnais (Mém. Soc. Acad. Boulogne-sur-Mer, vol. ii), 1867, p. 53.

Generic Characters.—Principal teeth elongated, irregularly quadrate, with slightly arched but flattened crown; symphysial teeth few and large, much arched, without lateral denticles, longitudinally keeled; all superficially marked by fine reticulate wrinkles or ridges. Dorsal fin-spines ornamented by stellate tubercles, sometimes in part fused into short longitudinal ribs; posterior denticles in two longitudinal series, often alternating, not marginal, but placed close together on a mesial ridge. One or two large hook-shaped dermal spines, each on a triradiate base, immediately behind the orbit, at least in males.

Type Species.—*Asteracanthus ornatissimus* (L. Agassiz, Poiss. Foss., vol. iii, 1837, p. 31, pl. viii), typically from the Upper Jurassic of Western Europe.

Remarks.—This genus is closely related to *Acrodus* and *Hybodus*, but the complete fish is unknown, and the teeth and spines have hitherto been found associated only in a variety of the type species discovered by Mr. Alfred N. Leeds in the Oxford Clay of Peterborough (A. S. Woodward, Ann. Mag. Nat. Hist. [6], vol. ii, 1888, pp. 336—342, pl. xii). Although dorsal fin-spines occur both in Wealden and in Purbeck formations, no teeth have yet been met with in the same horizons and localities.

1. Asteracanthus verrucosus, Egerton. Text-fig. 7.

1854. *Asteracanthus verrucosus*, P. M. G. Egerton, Ann. Mag. Nat. Hist. [2], vol. xiii, p. 433.
1855. *Asteracanthus verrucosus*, P. M. G. Egerton, Figs. and Descripts. Brit. Organic Remains (Mem. Geol. Surv.), dec. viii, pl. ii.
1889. *Asteracanthus verrucosus*, A. S. Woodward, Catal. Foss. Fishes, B. M., pt. i, p. 313.

FIG. 7.—*Asteracanthus verrucosus*, Egerton; dorsal fin-spine lacking posterior denticles, right side view, two-thirds nat. size.—Middle Purbeck Beds: Swanage, Dorset. Egerton Collection (B. M. no. P. 2209).

Type.—Dorsal fin-spine; Dorset County Museum, Dorchester.
Specific Characters.—Dorsal fin-spines attaining a maximum length of about

35 cm.; more or less gently arched and laterally compressed, but not keeled anteriorly; posterior face slightly raised, with denticles relatively smaller than in the type species; ornamental tubercles very numerous and closely arranged, mostly oval in form, and not only forming longitudinal series but also tending to arrangement in regular transverse series; the tubercles more or less fused into longitudinal ridges near the apex of the spine.

Description of Specimens.—The type spine is well preserved, lacking only the apex and the posterior denticles. The specimen shown in Text-fig. 7 is still finer, with the apex only a little worn and the posterior denticles again lacking. The rounded front border and the characteristic ornament are especially well seen. The posterior face is not sharply keeled, but only gently rounded, and in some other specimens in the British Museum its rather small denticles are arranged in two well-separated close series. The degree of curvature varies, and some spines are nearly straight, but all must have have been very obliquely inserted. It seems impossible at present to distinguish that of the anterior from that of the posterior dorsal fin.

Horizon and Locality.—Middle Purbeck Beds: Swanage, Dorset.

2. **Asteracanthus semiverrucosus,** Egerton. Text-fig. 8.

1854. *Asteracanthus semiverrucosus,* P. M. G. Egerton, Ann. Mag. Nat. Hist. [2], vol. xiii, p. 434.
1855. *Asteracanthus semiverrucosus,* P. M. G. Egerton, Figs. and Descripts. Brit. Organic Remains (Mem. Geol. Surv.), dec. viii, pl. iii.

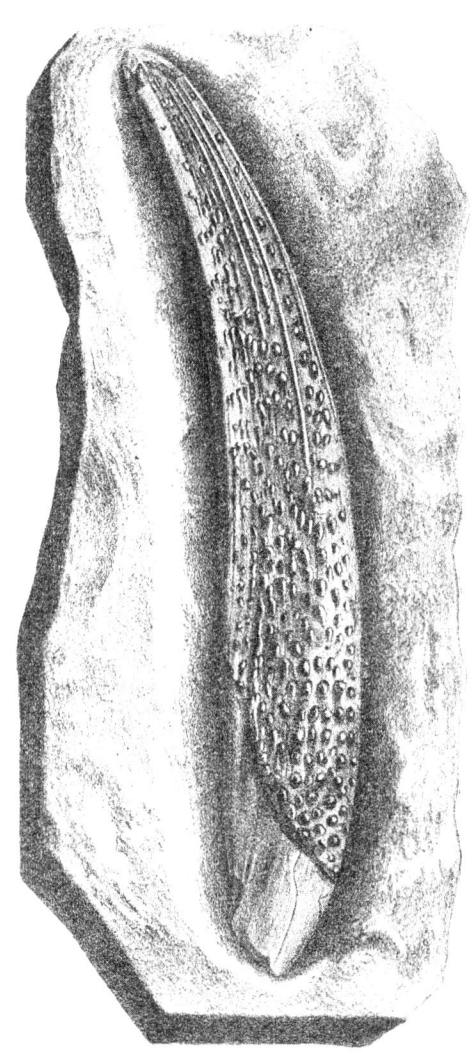

FIG. 8.—*Asteracanthus semiverrucosus,* Egerton; imperfect dorsal fin-spine, left side view, in matrix, two-thirds nat. size.—Middle Purbeck Beds: Swanage, Dorset. Dorset County Museum, Dorchester. After Egerton.

Type.—Imperfect dorsal fin-spine; Dorset County Museum, Dorchester.

Specific Characters.—Dorsal fin-spine about 25 cm. in length, much arched, laterally compressed and keeled anteriorly; ornamental tubercles ovate, very large, sparsely and rather irregularly arranged, some fused into longitudinal ribs.

Description of Specimen.—The type spine of this species still remains unique (Text-fig. 8). It lacks both the apex and much of the inserted base, but otherwise exhibits well its characters. The anterior tubercles in the basal half are the largest, those near the posterior border being relatively small and few. The fused ridges are in part slightly beaded. Some of the posterior denticles, preserved in the apical half, are rather large. The sharply arched form of the spine is especially noteworthy.

Horizon and Locality.—Middle Purbeck Beds: Swanage, Dorset.

3. **Asteracanthus granulosus,** Egerton. Text-fig. 9.

1854. *Asteracanthus granulosus,* P. M. G. Egerton, Ann. Mag. Nat. Hist. [2], vol. xiii, p. 433.
1855. *Asteracanthus granulosus,* P. M. G. Egerton, Figs. and Descripts. Brit. Organic Remains (Mem. Geol. Surv.), dec. viii, pl. i.
1859. *Asteracanthus granulosus,* Pictet and Campiche, Foss. Terr. Cretacé St. Croix, p. 98, pl. xii, fig. 11.

Type.—Dorsal fin-spine; British Museum.

Specific Characters.—Dorsal fin-spine nearly similar in form and proportions to that of *A. verrucosus,* but with the ornamental tubercles relatively smaller and less closely arranged.

Description of Specimens.—The type spine, although uncrushed, is a little abraded, incomplete at the apex, and without the posterior denticles. The front border is clearly shown to be rounded, and the posterior face is also only rounded, not keeled. The basal fragment of a second specimen described and figured by Egerton proves that the species sometimes attained a larger size than *A. verrucosus.* Another fine specimen, 30·5 cm. in length, found by Mr. Philip Rufford in the Wadhurst Clay at Ecclesbourne, near Hastings (Text-fig. 9), displays the side view a little fractured by crushing. The sparse ornamental tubercles tend to be arranged not only in longitudinal, but also sometimes in transverse series; while near the pointed tapering apex they are fused as usual into longitudinal ridges. The posterior denticles are seen to be relatively small.

Fig. 9.—*Asteracanthus granulosus,* Egerton; dorsal fin-spine, right side view, slightly more than one-half nat. size.—Wealden (Wadhurst Clay): Ecclesbourne, near Hastings, Sussex. Rufford Collection (B. M. no. P. 8939).

Horizon and Localities.—Wealden: Tilgate Forest, and Ecclesbourne, near Hastings, Sussex. Also in the Lower Neocomian of St. Croix, Switzerland.

Family MYLIOBATIDÆ (?).

Genus **HYLÆOBATIS**, novum.

Generic Characters.—Teeth more or less transversely elongated, with truncated ends, the crown overhanging the root on all borders. Oral surface of crown gently tumid, sloping down to its low and rounded anterior border, but sharply separated behind from a deep and concave posterior surface; covered with enamel which is variously marked with wrinkles.

Type Species.—*Hylæobatis problematica*, described below.

Remarks.—This genus is known only by a few small isolated teeth evidently of one species, and its affinities are uncertain. The teeth must have been arranged in a close tessellated pavement, as in the Cretaceous *Ptychodus* and the Tertiary Myliobatid skates. The slight bevelling of their ends shows that they alternated in transverse series; and occasional pressure-scars denote crowding. The superficial wrinkling of the enamel is rather suggestive of the rugosity round the margin of some teeth of *Ptychodus*; the deep concave posterior face of the crown and the low rounded anterior border also recall corresponding features in the same Cretaceous teeth. It may even be added that their microscopical structure (Text-fig. 10) agrees with that of the teeth of *Ptychodus* (Text-fig. 11) though there is nothing in this to distinguish them definitely from Cestraciont teeth. It is thus possible that *Hylæobatis* may prove to be one of the long-sought forerunners of *Ptychodus*.

1. **Hylæobatis problematica**, sp. nov. Plate V, figs. 1—5; Text-fig. 10.

Type.—Tooth without root; York Museum.

Specific Characters.—The type species, founded on isolated teeth measuring from 6 to 13 mm. in longest (transverse) diameter. Oral surface of dental crown feebly marked with coarse vermiculating wrinkles, which are more or less reticulate and pass at the front border into stronger vertical wrinkles; concave posterior face smooth or marked with slight vertical flutings; the anterior and posterior borders nearly straight and nearly parallel.

Description of Specimens.—The type tooth from Brook (Pl. V, fig. 1) lacks the root, but is otherwise well preserved. It is not bilaterally symmetrical, one end being wider and deeper than the other and distinctly bevelled for contact with two teeth. Its oral surface (fig. 1) is gently tumid, a little worn during life in its hinder half, but well preserved laterally and anteriorly, and feebly marked with irregular wrinkles which are directed chiefly along the longer (transverse) axis of the tooth. Seen from below (fig. 1 a) the crown clearly overhangs the root on all borders. In anterior view (fig. 1 b) the vertical wrinkling is conspicuous, and the crown is seen to be deepest at the wider end. In posterior view (fig. 1 c) the sharp margin of the oral surface forms a prominent ledge over the comparatively smooth, concave posterior face. The narrower end (fig. 1 d) is gently rounded, but marked by two well-separated small pressure-scars. The bevelled wider end (fig. 1 e) is also marked by two larger pressure-scars. A second tooth from Brook (Pl. V, fig. 2) is very strongly worn in its posterior half, and the worn surface is widest in the middle with a nearly semicircular margin. One end of the tooth is

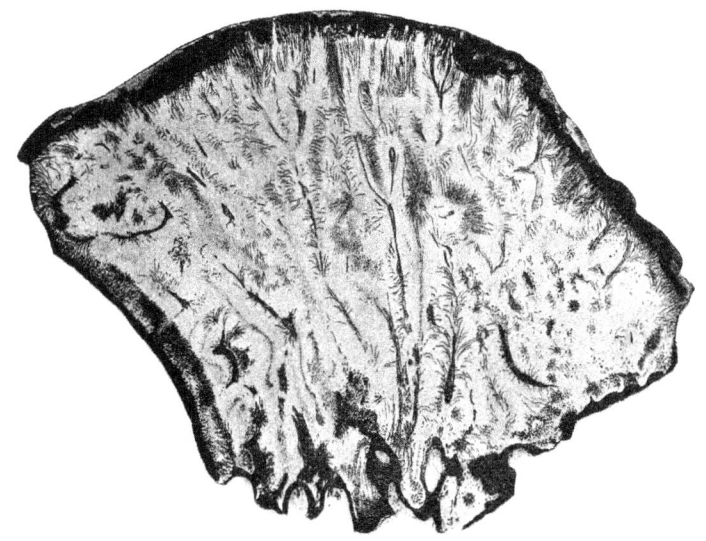

FIG. 10.—*Hylæobatis problematica*, gen. et sp. nov.; vertical antero-posterior section of crown of tooth, enlarged about 20 times.—Wealden: Sevenoaks, Kent. Sedgwick Museum, Cambridge.

again wider than the other and more distinctly bevelled for contact with two teeth; and its rounded anterior border (fig. 2 a) exhibits the vertical wrinkling. On the unworn part of the oral surface the irregular wrinkles are mainly in the direction of the long axis of the tooth. A larger and more transversely-elongated tooth from Sevenoaks (Pl. V, fig. 3) displays well the vertical wrinkling of its anterior border, but the oral surface seems to have been worn nearly smooth. One extensive pressure-scar occurs at each lateral end of the tooth, and the specimen is broken across to exhibit the transverse section (fig. 3 a). The microscopical structure of this transverse section is shown highly magnified in Text-fig. 10, where the darkly-stained enamel-layer (ganodentine) extends both over the upper oral surface and over the posterior concave surface, while the ordinary dentine is traversed by radiating and bifurcating vascular canals, which are bordered with canaliculi throughout their length and terminate in a tuft of canaliculi beneath the ganodentine.

A closely similar arrangement is seen in *Ptychodus* (Text-fig. 11). Another transversely-elongated tooth (Pl. V, fig. 5) has been so much worn during life that most of its surface markings have been removed; but it is interesting as showing a distinct bevelling at one end and a single well-marked pressure-scar at the other. The original of Pl. V, fig. 4, which is a less elongated tooth, has also been worn during life, but its oral face remains convex and seems to have been opposed to two teeth of a transverse series in the mouth. One lateral end of this tooth is strongly bevelled for articulation with two teeth.

Horizon and Localities.—Wealden: Brook, Isle of Wight; Sevenoaks, Kent.

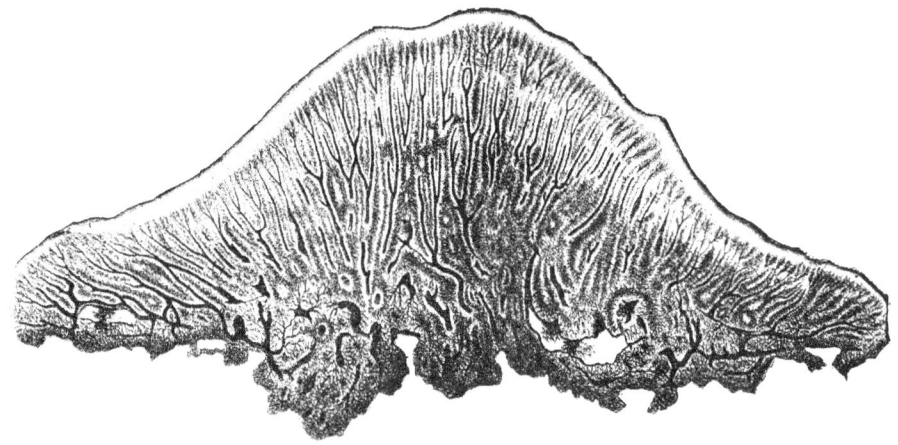

Fig. 11.—*Ptychodus mammillaris*, Agassiz; vertical transverse section of crown of tooth, highly magnified.—English Chalk. After Agassiz.

Subclass *TELEOSTOMI.*

Order *CROSSOPTERYGII.*

Family CŒLACANTHIDÆ.

Genus **UNDINA**, Münster.

Undina, G. von Münster, Neues Jahrb. f. Min., etc., 1834, p. 539.
Holophagus, P. M. G. Egerton, Figs. and Descripts. Brit. Organic Remains, dec. x (Mem. Geol. Surv., 1861), p. 19.

Generic Characters.—External bones and scales superficially ornamented with tubercles or fine interrupted ridges of ganoine; parafrontal and circumorbital bones plate-like, without superficial excavations. Teeth absent on the margin of the jaws, but a few hollow conical teeth within. Supplementary caudal fin prominent; the rays of all the fins robust, often expanded, and with numerous articulations in the distal portion; small upwardly-pointing denticles on the preaxial rays of the first dorsal and caudal fins.

Type Species.—*Undina penicillata* (Münster, *loc. cit.* and A. Wagner, Abhandl. math.-phys. Cl. k.-bay. Akad. Wiss., vol. ix, 1863, p. 696) from the Lithographic Stone (Lower Kimmeridgian) of Bavaria.

1. **Undina purbeckensis**, sp. nov. Plate IV, fig. 1.

Type.—Imperfect fish; British Museum.

Specific Characters.—A stout species attaining a length of at least 40 cm. Length of head with opercular apparatus about one quarter the total length of the fish, and somewhat less than the maximum depth of the trunk. Rays of first dorsal and caudal fins not expanded in the distal part, where the articulations are not very close; first dorsal fin with ten rays, caudal fin with about twenty rays above and below. Scales ornamented with coarse elongated tubercles, which are irregularly and rather sparsely arranged, shortest and most numerous on the dorsal scales, longest and fewest on the ventral scales.

Description of Specimen.—This species is known only by the single imperfect fish, shown of one half the natural size in Pl. IV, fig. 1. Of the head very little remains, but its shape and proportions are indicated by a fragment of the cranium at the bend of the frontal profile and by the nearly complete lower edge of the mandible (or perhaps gular plate) in cross-section. The hinder limit is marked by an imperfect impression of the clavicle. Part of the smooth outer face of the left pterygoquadrate bone is seen in position, and a displaced imperfect ceratohyal occurs beneath it. The neural arches of the vertebral axis of the trunk are well preserved, and about forty-five can be counted as far as the origin of the caudal fin, the hinder arches of the series gradually becoming longer and stouter. The ribs, seen only in the hinder half of the abdominal region, are small and delicate; but the hæmal arches in the tail are larger and more nearly symmetrical with the opposed neurals.

Of the paired fins, only one of the characteristic pelvic bones (*plv.*) remains, showing that the corresponding fin was inserted closer to the pectoral arch than to the tail. The ten stout rays of the first dorsal fin are distinct (d^1), but their laminar support is covered by the scales. No denticles are seen upon them, but there are shallow pits on the articulated distal portion of some rays indicating their original presence. The forked support for the second dorsal fin is preserved (d^2), and there are also some traces of the comparatively delicate rays. The anal fin is represented only by displaced fragments (*a*). The caudal fin, as usual, is preceded by two or three free supports above, probably also below; and its total number of rays cannot have been less than twenty above and below. The characteristic denticles are seen on some of the rays in the anterior part of the fin.

Many of the scales are sufficiently well preserved to exhibit their characteristic ornament. Near the dorsal border between the two fins some comparatively large scales bear numerous short and small elongated tubercles, which are very irregularly arranged (fig. 1 a). On the middle of the flank of the caudal region shorter and deeper scales are seen, with a somewhat coarser and sparser ornament of slightly elongated tubercles (fig. 1 c). At the ventral border, below the remains of the pelvic fins, the small and elongated scales are marked with closely arranged and much elongated tubercles (fig. 1 b).

The air-bladder is large, as usual, extending in the fossil from a point shortly behind the clavicle to the extreme posterior end of the abdominal region.

Affinities.—*Undina purbeckensis* appears to be closely related to the type species, *U. penicillata*, from the Lithographic Stone of Bavaria, but differs in being a stouter fish, with a finer tubercular ornament on the principal scales.

Horizon and Locality.—Middle Purbeck Beds: Swanage, Dorset.

Order ACTINOPTERYGII.

Family PALÆONISCIDÆ.

The latest known members of this family occur in the Purbeck and Wealden formations, and are referable to the highly specialised genus *Coccolepis*.

Genus **COCCOLEPIS**, Agassiz.

Coccolepis, L. Agassiz, Poiss. Foss., vol. ii, pt. i, 1844, p. 300.

Generic Characters.—Trunk elegantly fusiform. Mandibular suspensorium oblique; dentition consisting of an inner series of large laniaries flanked externally with minute teeth; external bones tuberculated or rugose. Fins large or of moderate size, the rays of all articulated and branching distally; fulcra minute or absent. Pelvic fins with extended base-line; dorsal and anal fins triangular, the former opposed to the space between the latter and the pelvic fins; upper caudal lobe much elongated, the fin deeply cleft and somewhat unsymmetrical. Scales thin and deeply imbricating, ornamented with tuberculations of ganoine.

Type Species.—*Coccolepis bucklandi* (L. Agassiz, Poiss. Foss., vol. ii, pt. i, 1844, p. 303, pl. xxxvi, figs 6, 7), from the Lithographic Stone (Lower Kimmeridgian) of Bavaria.

Remarks.—This genus ranges upwards from the Lower Lias of Lyme Regis (*Coccolepis liassica*, A. S. Woodward, Ann. Mag. Nat. Hist. [6], vol. v, 1890, p. 435, pl. xvi, figs. 2—4) to the Wealden of Bernissart, Belgium (*Coccolepis macropterus*, R. H. Traquair, Mém. Mus. Roy. Hist. Nat. Belg., vol. vi, 1911, p. 11, pl. i, text-figs. 1—3). It also has a wide geographical distribution, one species being known from the Jurassic of Talbragar, New South Wales (*Coccolepis australis*, A. S. Woodward, Mem. Geol. Surv. New South Wales, Palæont. no. 9, 1895, p. 5, pl. i; pl. ii, fig. 4; pl. v, fig. 1).

1. Coccolepis andrewsi, A. S. Woodward. Plate IV, figs. 2, 3.

1890. *Coccolepis andrewsi*, Woodward and Sherborn (*ex* Traquair, MS.), Cat. Brit. Foss. Vertebrata, p. 37 (name only).
1891. *Coccolepis andrewsi*, A. S. Woodward, Catal. Foss. Fishes, B. M., pt. ii, p. 524.
1895. *Coccolepis andrewsi*, A. S. Woodward, Geol. Mag. [4], vol. ii, p. 145, pl. vii, fig. 1.

Type.—Fish, wanting pectoral fins; Museum of Practical Geology, London.

Specific Characters.—A small species attaining a length of about 6 cm.: maximum depth of trunk contained about six times in the total length; upper caudal lobe excessively elongated and slender. Fin-rays smooth, with distant articulations. Dorsal fin arising somewhat in advance of the middle point of the back, partly opposed to the hinder portion of the pelvic fins, at least as deep as long, and its maximum depth nearly equalling that of the trunk at its point of origin; anal fin scarcely deeper than long, about two-thirds as extended as the dorsal, arising completely behind the latter and situated close to the caudal fin. Scales very coarsely granulated; fulcra of upper caudal lobe slender, much elongated, and very numerous.

Description of Specimens.—This species is known only by two specimens discovered by the Rev. W. R. Andrews, F.G.S., the one nearly complete (Pl. IV, fig. 2), the other showing the posterior abdominal and caudal regions (Pl. IV, figs. 3, 3 a).

The head of the type specimen appears to be typically Palæoniscid, but it is too imperfect for description. The only noteworthy features are a few slender conical teeth in the mandible, and traces of delicate broad branchiostegal rays below. The axial skeleton of the trunk is well exhibited through the thin squamation in both specimens, and is also typically Palæoniscid. The neural and hæmal arches, which bound the vacant space for the persistent notochord throughout the length of the fish, are only superficially ossified, appearing hollow in the fossilised state when broken. Their total number to the base of the caudal fin is about forty, and of these fifteen or sixteen may be reckoned as caudal. The neural spines in the abdominal region are stout and relatively large, not fused with their supporting arches; but both these and the hæmal spines are firmly fixed to their arches in the tail. There are no ribs, the hæmal arches in the abdominal region being merely a series of diminutive cartilages, best seen in the counterpart of the type specimen. At the beginning of the caudal region the hæmal arches suddenly become elongated, and five are distinct in advance of the anal fin in both specimens. Behind these there is some displacement of the hæmals in the type, but ten can be clearly counted in the second specimen as far as the origin of the caudal fin. The neural arches at the base of the upper caudal lobe are aborted, and a series of at least nine slender rods above them support the large fulcra. The hæmal arches in the basal part of the same lobe are enlarged for the direct support of the dermal rays; and the series is continued for some distance along the lobe by very small though stout ossified cartilages.

The pectoral fins are missing, but all the other fins are well preserved in both fossils. Ordinary fulcra are absent, but at the origin of each fin there are from three to six simple, though distantly articulated rays, gradually increasing in length to the apex of the fin, where the normal rays begin. These are also crossed by distant articulations, and, in the caudal fin at least, they are distally bifurcated. The pelvic fin is about as long as deep, arising nearly midway between the pectoral arch and the anal fin; its rays are shown to be not less than twenty in number, but the supports are unfortunately not observable. The number of rays in the dorsal fin is uncertain, but nineteen or twenty endoskeletal supports can be counted in the second specimen (fig. 3). The anal fin is somewhat smaller than the dorsal, with fourteen endoskeletal supports, of which the foremost is much the longest (fig. 3). The extreme elongation of the upper caudal lobe is best seen in the second specimen (fig. 3).

The whole of the trunk is covered with small, thin scales, which have the appearance of overlapping. They are, however, too obscure for detailed description, and it can only be noted that those of the lateral line in the caudal region are slightly thickened, and form a conspicuous smooth band along the flank as far as the beginning of the upper caudal lobe. Rather large tubercles of ganoine ornament the scales, and are especially well seen in parts of the abdominal region. A smooth thick ovate scale, pointed in front, occurs at the origin of the anal fin in both specimens (fig. 3 a). The oat-shaped scales on the slender caudal lobe are comparatively thick and smooth.

Horizon and Locality.—Lower Purbeck Beds: Teffont, Wiltshire.

2. **Coccolepis**, sp. Plate IV, fig. 4.

Coccolepis or a related Palæoniscid genus also occurs in the English Wealden, as shown by an imperfect maxilla discovered by Mr. Charles Dawson, F.G.S., in the Wadhurst Clay of Hastings (Pl. IV, fig. 4). The upper margin of the bone is incomplete, but the oral border is well preserved, and is seen to be bent sharply downwards behind. The outer face of the bone is smooth on the anterior extension, but very finely rugose in the hinder portion, where there is a tendency to delicate ridges concentric with the hinder and upper borders. The outer face is also traversed by a slight longitudinal groove inclined forwards and downwards. The minute outer teeth, so far as preserved, are very slender, but the large smooth conical teeth of the spaced inner series are tumid at the base.

Family SEMIONOTIDÆ.

Genus **LEPIDOTUS**, Agassiz.

Lepidotes, L. Agassiz, Jahrb. f. Min., Geogn., etc., 1832, p. 145.
Lepidotus, L. Agassiz, Poiss. Foss., vol. ii, pt. i, 1833, pp. 8, 233.
Sphærodus, L. Agassiz, tom. cit., pt. i, 1833, p. 15 (in part).
Scrobodus, G. von Münster, Neues Jahrb. f. Min., etc., 1842, p. 38.
Plesiodus, A. Wagner, Abh. k. bayer. Akad. Wiss., math.-phys. Cl., vol. xi, 1863, p. 632.
Prolepidotus, R. Michael, Zeitschr. deutsch. geol. Ges., vol. xlv, 1893, p. 729.

Generic Characters.—Trunk fusiform and only moderately compressed. Marginal teeth robust, styliform; inner teeth stouter, tritoral but smooth. Opercular apparatus well developed, with a narrow arched preoperculum, but with few branchiostegal rays and no gular plate. Ribs ossified. Fin-fulcra very large and biserial, present on all the fins. Paired fins small or of moderate size; dorsal and anal fins short and deep, the former just in advance of the latter; caudal fin slightly forked. Squamation regular and continuous, the scales rhombic, very robust, smooth or feebly ornamented; flank-scales not much deeper than broad, with their wide overlapped margin produced forwards at the upper and lower angles; scales of dorsal and ventral aspect nearly as deep as broad; dorsal and ventral ridge-scales usually inconspicuous.

Type Species.—*Lepidotus elvensis* (*Cyprinus elvensis*, H. D. de Blainville, Nouv. Dict. d'Hist. Nat., vol. xxvii, 1818, p. 394; *Lepidotus gigas*, L. Agassiz, Poiss. Foss., vol. ii, pt. i, 1833—37, pp. 8, 235, pls. xxviii, xxix) from the Upper Lias of France, Würtemberg, Bavaria, and England. It is described in detail by F. A. Quenstedt, " Ueber *Lepidotus* im Lias ε" (Tübingen, 1847); and the French type specimen in the National Museum of Natural History, Paris, is described and figured by F. Priem, Annales de Paléontologie, vol. iii (1908), p. 5, pl. ii. A closely related species, *Lepidotus semiserratus,* Agassiz, is described by A. S. Woodward, Proc. Yorks. Geol. and Polyt. Soc., n. s., vol. xiii, 1897, pp. 325—336, pls. xlvi—xlviii.

On comparing the head of one of the earlier species of *Lepidotus* (Text-fig. 12) with that of one of the latest (Text-fig. 13), it will be noticed that as the teeth become stouter, the jaws are shortened and the mouth is relatively smaller. The supratemporals and the cheek-plates become more or less irregularly subdivided; and the wavy median suture between the parietal and frontal bones of the cranial roof becomes nearly straight.

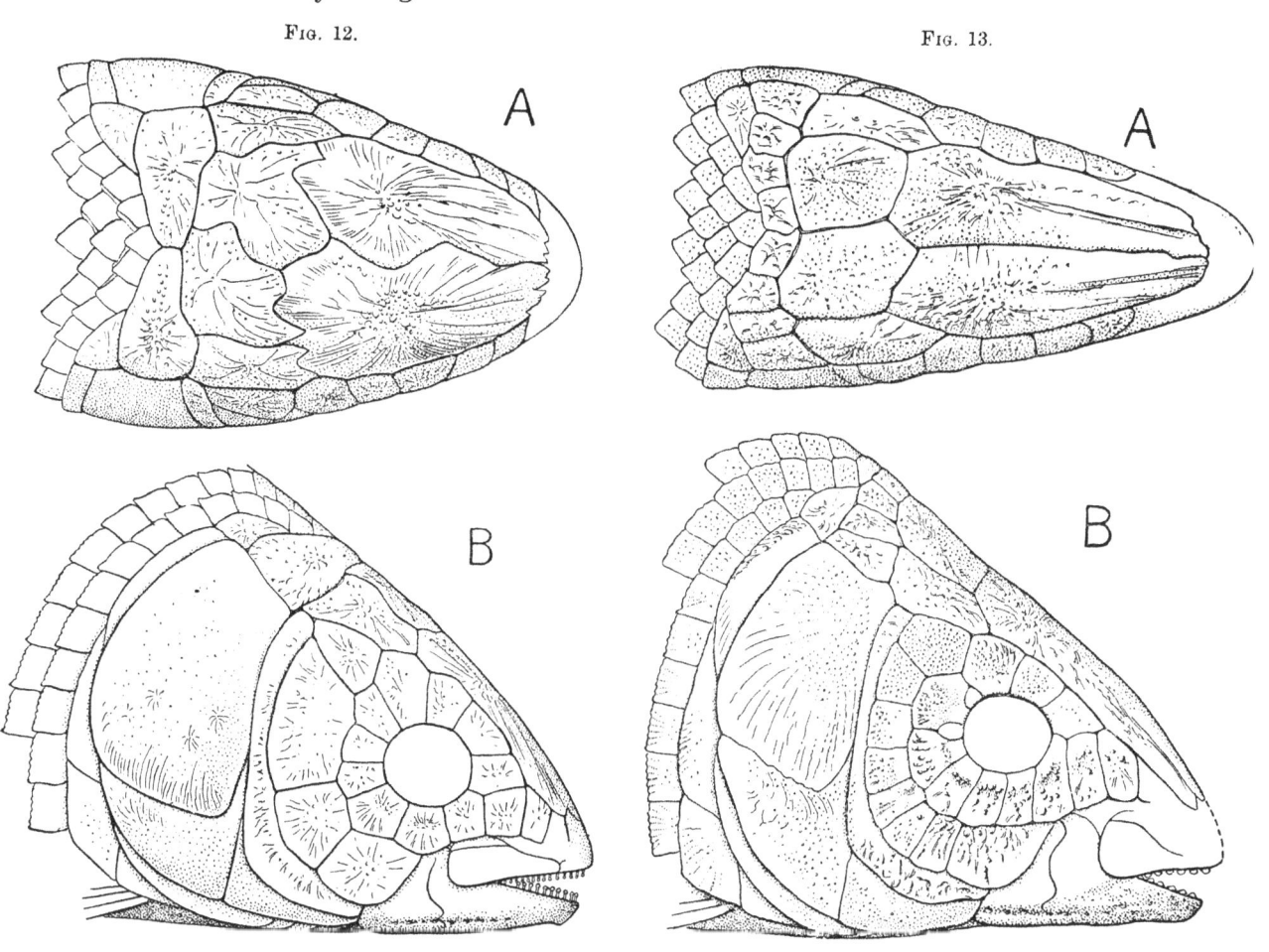

Fig. 12.—*Lepidotus semiserratus*, Agassiz; restoration of head, upper (A) and right side view (B), one-half nat. size.—Upper Lias: Whitby, Yorkshire.

Fig. 13.—*Lepidotus mantelli*, Agassiz; restoration of head, upper (A) and right side view (B), one-half nat. size.—Wealden: Sussex.

1. **Lepidotus minor**, Agassiz. Plate V, figs. 6—11; Plate VI; Plate VII, figs. 1—5; Text-figure 14.

1833–37. *Lepidotus minor*, L. Agassiz, Poiss. Foss., vol. ii, pt. i, pp. 9, 260, pl. xxxiv (*non* pl. xxix c, fig. 12).

1887. *Lepidotus minor*, W. Branco, Abh. geol. Specialk. Preussen u. Thüring. Staaten, vol. vii, p. 363, pl. vi, fig. 2 (? *non* p. 366, pl. vi, fig. 1).

1893. *Lepidotus minor*, A. S. Woodward, Proc. Zool. Soc., p. 562, pl. xlix, fig. 3.

1895. *Lepidotus minor*, A. S. Woodward, Catal. Foss. Fishes, Brit. Mus., pt. iii, p. 94, text-fig. 22.

Type.—Imperfect fish; School of Mines, Paris.

Specific Characters.—A species attaining a length of about 40 cm., but usually less. Length of head with opercular apparatus exceeding three-quarters the maximum depth of the trunk, and slightly less than one-quarter the total length of the fish. Frontal profile steep and snout acute; parietal bones about one third as long as the frontals, which are four times as long as their maximum width, narrow in front, and united by a nearly straight median suture; two postorbital plates, the lower large and much deeper than wide, and both wider than the circumorbitals; mandibular symphysis not much deepened or extended; external bones rugose and more or less tuberculated, except the maxilla and premaxilla, which are smooth. A single pair of supratemporal plates. Teeth on moderately long pedicles, constricted below the crown, and all with a small pointed apex when unworn. Operculum about three-fifths as wide as deep, its width contained three times in the length of the head. Fin-fulcra very large, the principal dorsal fulcra

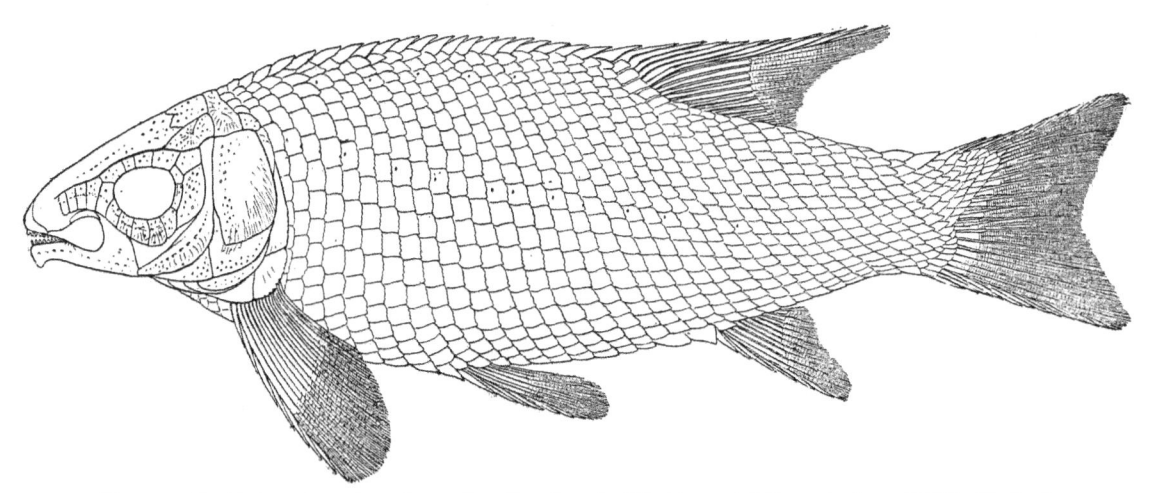

FIG. 14.—*Lepidotus minor*, Agassiz; restoration, from half to three-eighths nat. size.—Middle Purbeck Beds: Swanage, Dorset.

more than half as long as the anterior dorsal fin-rays, and four to six directly inserted in the ridge of the back; pelvic fins arising midway between the pectorals and the anal; dorsal fin much larger than the anal, but each with ten or eleven rays. Scales smooth, usually more less serrated on the flank; scales of lateral line notched; dorsal ridge-scales acutely pointed and rather prominent.

Description of Specimens.—This species varies considerably in the shape of the trunk, and the fossils are often distorted by crushing. The degree of serration of the flank-scales and the relative size of the fin-fulcra are also variable. Every gradation, however, can be found between the deepened form with rounded back, shown in the type specimen and in our Pl. VI, and the more slender type which is figured by Branco (*loc. cit.*, 1887), and represented here in Text-fig. 14. A careful study of a large series of specimens also seems to show that the nature of the serrations of the scales is not correlated either with the size of the fish or with its shape.

The chondrocranium is unknown, but it must have been imperfectly ossified, and, as shown by specimens in the British Museum (Pl. V, fig. 6) and the Warwick Museum, the membrane bones of the roof were readily detachable from it. The parietal bones (Pl. V, figs. 6, 7, *pa.*) form a nearly symmetrical pair united in a slightly wavy median suture; and that of one side extends a little further forwards on the cranial roof than that of the other side. Each bone is about one and a half times as long as broad, and its outer face is not only rugose, but also coarsely and irregularly tuberculated, and marked in the middle by a short transverse groove for the slime-canal. The squamosal on each side is a longer and narrower bone, united in a slightly wavy suture with the parietal, extending from the occiput behind to the postorbital prominence of the frontal forwards. Its outer face (Pl.V, figs. 7, 8, *sq.*) is also rugose and coarsely tuberculated, while its inner face (Pl. V, fig. 6, *sq.*) bears the usual elongated articular facette for the upper end of the hyomandibular. The frontals (Pl. V, figs. 6—8, *fr.*) are at least two and a half times as long as the parietals, and are also united in a slightly wavy median suture. Each is widest behind, the maximum width being about a quarter of the total length; and its comparatively narrow anterior portion ends in a few pointed digitations. Its outer margin is regularly excavated for the relatively large orbit, thus producing definite postorbital and preorbital prominences. Its outer face is coarsely and irregularly tuberculated as far forwards as the middle of the orbit, while the anterior portion is marked only with a few irregular longitudinal grooves. The longitudinal slime-canal, which opens on the outer face in rather large pores, is enclosed by bone which forms a rounded ridge on the inner face (Pl. V, fig. 6, *fr.*). The ethmoid region and the nasals remain unknown. The parasphenoid (Pl. V, fig. 9) which is commonly ascribed to this species but has not yet been definitely seen in position, closely resembles that of *Lepidotus latifrons*.[1] It is narrowest at the small digitate basipterygoid processes, expanding much behind and exhibiting a deep cleft in its hinder margin. It is pierced between the basipterygoid processes by a foramen for the passage of the internal carotids, and it is toothless.

The cheek is completely covered with plates, which form a circumorbital ring, bounded behind by postorbitals and continued in front by a few preorbitals. All are irregularly tuberculated, while the circumorbitals behind and below the orbit, as well as the preorbitals, are also marked by the usual slime-canal. Of the circumorbital ring, four plates occur above the eye, occupying the excavation in the frontal border between the preorbital and postorbital prominences and extending backwards to bound the anterior end of the squamosal (see Pl. V, figs. 7, 8). These plates are variable in shape, but the foremost is usually the longest. The two posterior circumorbitals are comparatively small and narrow, while the lower

[1] A. S. Woodward, 'On the Cranial Osteology of the Mesozoic Ganoid Fishes, *Lepidotus* and *Dapedius*.' Proc. Zool. Soc., 1893, p. 561, text-fig. 3.

of the two is also deep (see Pl. V, figs. 6, 8); and the five lower circumorbitals (*co.*) are relatively large and deep, followed in front by the still deeper and narrower plates of the preorbital series (Pl. V, fig. 8, and Pl. VI, *pro.*). The upper plate of the postorbital series (best seen in Pl. V, fig. 7, *po.*), articulating with the middle part of the squamosal border, is in contact with both the posterior upper and the upper posterior circumorbital plates, and bounds above the relatively large principal postorbital plate. This element (Pl. V, figs. 6—8; Pl. VI; *po.*) is about twice as deep as broad, of irregularly rhomboidal form, with the postero-inferior angle rounded, and its outer face sparsely ornamented by tubercles of ganoine, which tend to elongate by irregular fusion. There is no clear evidence of other cheek-plates of the outer or postorbital series; but in the original of Pl. VI, and in another specimen in the Museum of Sherborne School, there are fragments which may perhaps be interpreted as representing comparatively thin plates below the large postorbital continuing this series downwards and forwards.

The mandibular suspensorium is inclined forwards so that the quadrate articulation is beneath the middle of the orbit. The hyomandibular (Pl. V, fig. 10) is a thin lamina of bone, slightly more than twice as deep as wide, and strengthened on its outer face by a longitudinal ridge, from which a cross-ridge arises at the level of the prominence for the support of the operculum. The quadrate, not yet clearly seen, must also have been a delicate bone, while the entopterygoid and metapterygoid are comparatively thin laminæ (as shown in B. M. no. P. 5591). The ectopterygoid, however, is stout and horizontally expanded for the support of at least three rows of teeth parallel with the outer border. Its outer face (Pl. VII, fig. 1 *a*), which is conspicuous in several specimens, is smooth and flattened or very slightly concave, appearing as a narrow band, with the oral border gently concave and the upper anterior and posterior angles rounded off. This face is mistaken for that of the maxilla in the restoration published in the 'Catalogue of Fossil Fishes in the British Museum,' pt. iii (1895), p. 95, fig. 22. The horizontal extension of the ectopterygoid (Pl. VII, fig. 1) is widest in front, where its bevelled end passes under the hinder margin of the equally expanded but relatively small and thin palatine element (well seen in the specimen in Sherborne School Museum). Its clustered teeth, as also those of the imperfectly known vomer, are relatively larger and stouter than the ectopterygoid teeth. The maxilla (Pl. VI, *mx.*) is a delicate smooth lamina of bone, deepest and rounded behind, tapering forwards and terminating in front in a slender, inwardly directed process for articulation with the palatine. Its oral border, at least in the middle and anterior portion, bears a spaced series of comparatively small and slender teeth (some seen in B. M. no. 42308); while the upper border of its hinder expansion is capped by a single long and narrow supramaxilla (seen in Pl. VI). The premaxilla (Pl. V, fig. 11), seen in position in Pl. V, fig. 7, and in Pl. VI (*pmx.*), is a stouter smooth bone occupying only about half as much of the oral border as the maxilla, but bearing a regular spaced series of six or seven styliform teeth. Its hinder half tapers backwards to a blunt point, which meets the maxilla; its anterior half is continued upwards in a narrow laminar process, which is so long that (as in *Amia*) it passes beneath the nasal bone to articulate in a squamous suture with the inner face of the frontal (Pl. V, fig. 6, *pmx.*). This process, also as in *Amia*, is hollowed on its superior face and pierced by a large oval vacuity. The mandible is short and much deepened in the coronoid region. As seen from the outer face (Pl. VI, *md.*) the angular bone is comparatively short and deep, and marked near its lower margin by a row of large pits and rugosities for the slime-canal. The dentary bone (Pl. VII, fig. 2) is still deeper in the coronoid region just in front of the angular, but rapidly contracts to the tooth-bearing portion, which forms a narrow bar curving inwards and slightly deepening at the symphysis. This portion bears a regular spaced series of nine or ten styliform teeth somewhat larger than those of the upper jaw; while its outer face is a little rugose and bears a row of large pits for the course of the slime-canal. The summit of the coronoid region is formed by a long and narrow coronoid bone, exactly as in *Amia* (seen in Pl. VI). The articular end of the meckelian cartilage is slightly ossified (seen in B. M. no. 41157), and on the inner face of the mandible only one splenial element has been observed. This is comparatively stout, enters the mandibular symphysis, and bears a cluster of about three rows of teeth which are stouter than those of the dentary and diminish to comparatively small teeth behind (Pl. V, fig. 6, *md.*). All the teeth are hollow and fused with the supporting bone, not in sockets. When unworn the smooth and rounded enamelled crown rises to a sharp median apex, while the comparatively long pedicle is slightly swollen just below the crown and a little expanded at its base, which sometimes exhibits short vertical grooves. Successional teeth have been observed as in other species of *Lepidotus*.

In the hyoid arch the ceratohyal is relatively large, laterally compressed, constricted in the middle, and deepest behind (Pl. V, fig. 8, *ch.*). It does not appear to bear any branchiostegal rays.

The preoperculum, best seen in Pl. V, fig. 8, *pop.*, but also shown in Pl. V, fig. 6, and Pl. VI, is narrow and gently curved at the angle. The ascending limb is slightly constricted in its lower half, but expands upwards where its outer face is smooth and its truncated end is in contact with the squamosal bone. Its curved lower limb is more expanded and traversed by the usual longitudinal ridge, behind and below which the large openings of the slime-canal are conspicuous. The anterior border exhibits a smooth overlapped surface for the cheek-plates as far as this ridge; while the expansion behind, when unabraded, is marked by a few radiating crimpings, and the posterior edge usually bears a spaced row of large tubercles of ganoine. The operculum, also best seen in Pl. V, fig. 8, *op.*, where the postero-superior angle alone is incomplete, is somewhat wider below than above, and its maximum width measures about three-fifths of its depth. Its upper and anterior portions are sparsely ornamented with tubercles, which fuse together more or less irregularly and rarely spread over the whole plate. The suboperculum is usually about one-quarter as deep as the operculum, with a relatively large ascending process in front, and sparsely tuberculated near the anterior and inferior margins (Pl. V, fig. 8, *sop.*; Pl. VI, *sop.*). The interoperculum is small, elongate-triangular in shape, also sparsely ornamented with large irregular tubercles (Pl. V, fig. 8, *iop.*; Pl. VI, *iop.*). Only four branchiostegal rays have been clearly seen (Pl. V, fig. 8, *br.*), though there appear to be traces of two or three more in the original of Pl. VI. The uppermost is largest and bears a few tubercles, while the others are smooth. There is no gular plate.

The notochord must have been persistent, and there are no surrounding ossifications. The neural and hæmal arches are incompletely ossified, so that they appear hollow and are frequently crushed in the fossils. The neural spines in the abdominal region so far back as the origin of the dorsal fin, are stout and a little widened distally where they nearly reach the dorsal border (Pl. VII, fig. 3, *n*): they appear to be slightly curved, with the concavity forwards. The neural spines in the caudal region are shorter and more slender. The ribs (*r.*) are round or ovoid in section, comparatively slender, and extend almost to the ventral border. The hæmal arches within the base of the caudal fin are especially stout and somewhat expanded distally (as seen in B. M. no. P. 4989 *a*).

A single pair of large supratemporal plates (Pl. V, fig. 7, *st.*) overlaps the occiput, exhibiting the same rugosity and sparse tuberculation as the hinder bones of the cranial roof, and marked by the usual groove for the transverse slime-canal. Each supratemporal is wider than long and tapers towards the middle line of the fish, where it meets its fellow of the opposite side. The comparatively small exposed face of the post-temporal (Pl. V, fig. 7, *ptt.*) is triangular in shape and also tuberculated. Its inner face is smooth, and in the original of Pl. V, fig. 6, *ptt.*, shows no feature beyond the articular facette for the supraclavicle; but beneath the same bone in B. M. no. P. 5591, there is a displaced long slender process which seems to correspond with the internal descending process of the post-temporal in *Amia*. Such a process has already been discovered by Mr. Alfred N. Leeds on a post-temporal of *Lepidotus* from the Oxford Clay of Peterborough. The supraclavicle is a deep and narrow plate of bone, only rugose at its upper end where it is crossed obliquely by the slime-canal, and not serrated along its posterior edge. The clavicle curves well forwards and its large expanded lower end is seen from within below and just behind the mandible in Pl. VI. When the concave anterior margin is exposed (as in the Sherborne School specimen), it is seen to be covered with short oblique rows of small granulations. There are three large enamelled post-clavicular plates, of which the upper two are shown in Pl. V, fig. 8, *pcl.*, and in Pl. VI. They are smooth and exhibit a variable amount of serration of the posterior edge in the lower portion. The upper postclavicular is deep and narrow, and its tapering upper end is in contact with the lower part of the supraclavicle. The second plate, directly beneath the first, is not much deeper than wide, with a gently rounded postero-inferior angle; the third plate (not seen in the specimens figured, but shown in Text-fig. 14) is directly in front of the second plate, comparatively small and triangular in shape. The pectoral fin, as shown in Pl. VI, is rather large, the longest of its sixteen or seventeen closely pressed rays almost reaching the origin of the pelvic fins. The fin is fringed in front with a paired series of about eight enamelled fulcra, and all the rays, which are almost without enamel, are articulated and divided in their distal half. The pelvic fins nearly resemble the pectorals, but are much smaller, with only seven rays, of which the longest attains about three-quarters of the length of the longest pectoral. The dorsal and anal fins are characterised by their very stout fulcra, which are covered with smooth enamel and paired like those of the pectoral and pelvic fins. In the dorsal fin, four, five, or even six stout fulcra of gradually increasing length are directly inserted in the ridge of the back, and their long crowded supports (causing a break in the squamation in Pl. VI) penetrate the muscles almost as far as the position of the notochord. As shown in Pl. VI, the length of the longest of these fulcra much exceeds half the extreme height of the fin; while there are also a few large fulcra fringing the foremost ray. The rays are about eleven or twelve in number, closely arranged and rapidly diminishing in size backwards, each articulated and finely divided in the distal half: they are destitute of enamel or marked only by the slightest streak. The anal fin, with ten rays, nearly resembles the dorsal in shape, but it is much smaller and its anterior fulcra are less stout, only two or at most three being directly inserted in the ridge of the body. The great support for the fulcra, however, is deeply inserted in the muscles of the trunk (Pl. VII, fig. 5 *a*, *f.s.*). The caudal fin consists of about eighteen comparatively stout rays, which are closely articulated and divided nearly to the base, and differ from the other fin-rays in being well covered with enamel in their basal half. The fin is slightly excavated behind (as shown in B. M. no. 19006), and its fringing fulcra are about as large as those of the anal fin.

All the scales are smooth, sometimes with a faintly concave face; and their hinder margin is often slightly convex, while their upper and lower margins are sigmoidally curved. None of the principal flank-scales are much deeper than broad, but they are comparatively deeper in the typical stout forms (Pl. VI) than in the more slender forms (Text-fig. 14). They also appear sometimes smaller in the former than in the latter, but the total number of transverse series of scales is always approximately the same (about thirty-eight to forty as counted along the lateral line), while the number in a series above the origin of the pelvic fin is nineteen or twenty. In this position the lateral line always traverses the eleventh scale from the ventral border; while above the origin of the anal fin it traverses the thirteenth or fourteenth scale in the stout forms (Pl. VI), the ninth scale in the more slender forms (Text-fig. 14). The flank-scales in the abdominal region are nearly always feebly and coarsely serrated, but the serrations are usually restricted to the lower portion of each posterior margin; the scales of the lateral line are also often notched at the posterior margin, and a varying number are pierced by a simple foramen for the passage of the slime-canal. As the serrations of the flank-scales disappear towards the caudal region, the postero-inferior angle of each scale tends to become produced into a slender point, and this may even be strengthened by a slight ridge (B. M. no. 19006). Towards the dorsal and ventral borders, and in the hinder part of the caudal region, the scales become slightly broader than deep; and the third row from the dorsal border, between the occiput and the origin of the dorsal fin, is marked at intervals with the orifices of an upper slime-canal. The dorsal ridge-scales from a point shortly behind the occiput to the origin of the dorsal fin (accidentally removed from the original of Pl. VI, but seen in Pl. VII, fig. 4, and in Text-fig. 14), are relatively large and pointed, and form a conspicuous imbricated row; there are also two or three enlarged ridge-scales, not so deeply imbricating, at the origin of the caudal fin above and below. The three enlarged scales, with denticulated posterior margin, surrounding the anus just in front of the anal fin, are well shown somewhat displaced in Pl. VI. At the base of the fulcra of the dorsal fin about four scales in a regular row are elongated in the direction of the fin-rays; at the base of the caudal fin small elongated scales are similarly related to the rays, while a notch in the end of the caudal pedicle marks the limit of the triangular remnant of the upper caudal lobe.

The scales of the abdominal region are united in the usual manner by a complicated overlap and a peg-and-socket articulation. On the flank, the wide overlapped margin of each scale is produced at its upper angle into a large bluntly rounded process, at its lower angle into a smaller acute process (Pl. VII, fig. 5); while there is sometimes a minute prominence in the excavation between the two (*e. g.* in B. M. no. P. 4989 *a*). In the ventral scales (Pl. VII, fig. 5 *b*), the antero-superior process becomes especially large, produced upwards as well as forwards, and the antero-inferior process disappears. In all these scales, the peg-and-socket articulation, strengthened by a vertical ridge, is also present; in the caudal scales only the strengthening ridge, more or less widened, remains.

Horizon and Locality.—Middle Purbeck Beds: Swanage, Dorset.

2. Lepidotus notopterus, Agassiz. Plate VII, fig. 6.

1835-37. *Lepidotus notopterus*, L. Agassiz, Poiss. Foss., vol. ii, pt. i, p. 257, pl. xxxv.
1850. *Lepidotus notopterus ?*, V. Thiollière, Ann. Sci. Phys. Nat. Lyon [2], vol. iii, p. 138.
1852. *Lepidotus notopterus*, F. A. Quenstedt, Handb. Petrefakt., p. 197, pl. xv, fig. 4.
1863. *Lepidotus notopterus*, A. Wagner, Abhandl. k. bayer. Akad. Wiss., math.-phys. Cl., vol. ix, p. 628.

1873. *Lepidotus notopterus* ?, V. Thiollière, Poiss. Foss. Bugey, pt. ii, p. 15, pl. iv.
1887. *Lepidotus notopterus*, W. Branco, Abhandl. geol. Specialk. Preussen u. Thüring. Staaten, vol. vii, p. 382, pl. viii, fig. 5.
1887. *Lepidotus notopterus*, K. A. von Zittel, Handb. Palæont., vol. iii, p. 209, fig. 218.
1895. *Lepidotus notopterus*, A. S. Woodward, Catal. Foss. Fishes, Brit. Mus., pt. iii, p. 92.

Type.—Imperfect fish; British Museum.

Specific Characters.—A species attaining a length of about 40 cm. Length of head with opercular apparatus nearly equal to the maximum depth of the trunk, and occupying about one-quarter the total length of the fish. Snout acute; cranial roof-bones with few sparse tuberculations; teeth on moderately long pedicles. Operculum twice as deep as its maximum breadth, which is contained at least three times in the length of the head. Fin-fulcra very large, the principal dorsal fulcra sometimes half as long as the anterior dorsal fin-rays, and three or four directly inserted in the ridge of the back; the pelvic fins arising much nearer to the anal than to the pectorals; dorsal and anal fins deeper than long, the former larger than the latter. Scales smooth, very few serrated, but those of the lateral line and sometimes a few anterior flank-scales slightly notched on the hinder margin.

Description of Specimen.—An imperfect fish lacking most of the head (Pl. VII, fig. 6), obtained by the late Earl of Enniskillen from the Purbeck Beds of Swanage, differs from *L. minor* in its smaller and more delicate dorsal fin-fulcra, and agrees well, so far as it can be compared, with *L. notopterus* from the Lithographic Stone (Lower Kimmeridgian) of Germany and France. It may therefore be recorded provisionally under the latter specific name.

Traces of unusually stout ribs are exposed by the removal of the scales in the anterior part of the abdominal region. On the rest of the trunk the scales are in regular order, and the outline of the fish is only marred by the accidental removal of the ventral part of the abdominal region and by slight crushing in front of the position of the anal fin. The principal flank-scales in the abdominal region are slightly deeper than wide, with a smooth and somewhat convex hinder margin; above and below they are more nearly equilateral; and the dorsal ridge-scales are acuminate, though not enlarged. The scales in the caudal region are less deep, with a tendency towards the production of the postero-inferior angle; some of the ventral scales are much wider than deep; and two of the acuminate dorsal ridge-scales at the origin of the upper caudal lobe are a little enlarged. The scales of the lateral line are pierced at irregular intervals with large pores, and the hinder margin of each scale is notched near the postero-inferior angle. In the dorsal fin only eight rays can be counted, rapidly decreasing in length backwards, where two or three may be missing; the fulcra are very slender, the longest scarcely exceeding one-third of the length of the anterior fin-ray, and not more than three being directly inserted in the ridge of the back. The position of the anal fin, behind the dorsal, is indicated by traces of fulcra and the bases of fin-rays. The end of the caudal pedicle is deeply excavated at the base of the caudal fin, and the stout upper lobe is considerably produced. The caudal fin-rays exhibit the usual stoutness and close articulation, while the fulcra on the lower margin are large though slender.

Horizon and Locality.—Middle Purbeck Beds: Swanage, Dorset.

3. **Lepidotus mantelli**, Agassiz. (Plate VII, fig. 7; Plates VIII, IX, X; Plate XI, figs. 1—14; Text-figs. 13, 16—18.)

1826. Figure of dentary by T. Webster, Trans. Geol. Soc. [2], vol. ii, pl. vi, figs. 5, 6.
1827. "Scales of a Quadrangular Form," G. A. Mantell, Illustr. Geol. Sussex and Foss. Tilgate Forest, p. 58, pl. v, figs. 3, 4, 15, 16.
1833. *Lepidotus subdenticulatus*, L. Agassiz, Poiss. Foss., vol. ii, pt. i, p. 9. [Scales, afterwards referred to *L. fittoni*, Agassiz, tom. cit., p. 265; National Museum of Natural History, Paris.]
1833-37. *Lepidotus mantelli*, L. Agassiz, Poiss. Foss., vol. ii, pt. i, pp. 9, 262, pl. xxx, figs. 10—15; pl. xxx *a*, figs. 4—6; pl. xxx *b*, fig. 2; pl. xxx *c*, figs. 1—7.
1834-44. *Lepidotus fittoni*, L. Agassiz, Poiss. Foss., vol. ii, pt. i, p. 265, pl. xxx, figs. 4—6; pl. xxx *a* (excl. figs. 4—6); pl. xxx *b* (excl. fig. 2). [Portion of fish, British Museum.]
1836-44. *Tetragonolepis mastodonteus*, L. Agassiz, Poiss. Foss., vol. ii, pt. i, p. 216, pl. xxiii *e*, figs. 3, 4 (*non* fig. 5). [Small dentary.]
1841. *Lepidotus mantelli*, R. Owen, Odontogr., p. 69, pl. xxx, fig. 1; pl. xxxi. [Structure of teeth.]
1849. *Lepidotus mantelli*, W. C. Williamson, Phil. Trans., p. 444. [Structure of scales.]
1860. *Lepidotus fittoni*, J. E. Lee, Geologist, vol. iii, p. 458, pl. xii. [Structure of scales.]
1887. *Lepidotus mantelli*, W. Branco, Abhandl. geol. Specialk. Preussen u. Thüring. Staaten, vol. vii, p. 345, pl. iii, figs. 1, 2.
1895. *Lepidotus mantelli*, A. S. Woodward, Catal. Foss. Fishes, Brit. Mus., pt. iii, p. 108, text-figs. 23, 24.

Type.—Portion of fish; British Museum.

Specific Characters.—A stout species attaining a length of about 1 metre. Length of head with opercular apparatus considerably less than the maximum depth of the trunk and contained slightly more than four times in the total length of the fish. Snout acute and frontal profile somewhat bent; external bones more or less rugose or tuberculated; parietal bones much less than half as long as the frontals, which are about three times as long as their maximum width, very narrow in front, and united throughout their length by a nearly straight median suture; orbit unusually small, with a relatively large circumorbital ring, and the postorbital plates much subdivided, the lowest and foremost plate of this series being the largest. Mouth small, the mandibular articulation below the middle of the orbit; maxilla smooth, with deep rounded expansion behind; mandibular symphysis very robust, the dentary being much horizontally extended to support the large tooth-bearing splenial. Inner teeth very short and stout, smooth, usually with slightly acuminate crown when unworn; marginal teeth also stout, smooth and acuminate. Maximum width of operculum nearly two-thirds as great as its depth, and equalling about one-third the length of the head. Ring-vertebræ present in the adult. Fin-fulcra very large; pelvic fins much smaller than the pectorals and inserted nearer to the latter than to the anal fin; dorsal and anal fins almost equally elevated, with about fourteen and ten rays respectively, and the former fin with four or five fulcra directly inserted in the back; anal fin arising opposite hinder end of dorsal. Scales smooth, but showing coarse oblique corrugations when abraded, and those on the flank more or less finely serrated; principal flank-scales somewhat deeper than broad, those of the lateral line notched; dorsal ridge-scales acuminate, but usually inconspicuous.

Description of Specimens.—The type specimen of *Lepidotus mantelli* is the hinder portion of the head, with a fragment of the abdominal region and the base of a pectoral fin (Pl. VIII, fig. 1), in the Mantell Collection. The type specimen of the so-called *L. fittoni* is a vertically crushed and much abraded head, with part of the abdominal region (Pl. VIII, figs. 2, 2 *a*, 2 *b*). There are notable differences between these two fossils, as already pointed out by Agassiz; but the large collection in the British Museum seems to show that these differences are due partly to crushing, partly to abrasion, and partly to great variation in one and the same species. The earliest form is represented by a small specimen from the Purbeck Beds of Netherfield, Sussex, in the Hastings Museum (Pl. X, fig. 3).

The general shape and proportions of the fish are best shown in a specimen discovered by Mr. Charles Dawson in the Wadhurst Clay near Hastings (Pl. VII, fig. 7). This fossil is almost uncrushed, only slightly bent sideways at the base of the caudal fin, and the gently arched contour of the back is especially well displayed. The maximum depth of the trunk is somewhat less than a third of the total length of the fish, while the head must have occupied nearly a quarter of the same length. The greater part of the caudal fin is, of course, missing in this specimen.

The chondrocranium is well ossified, the various elements appearing in the fossil as pieces of thick, spongy bone. No supraoccipital has been observed, but its ordinary place in the occiput is occupied by the inner part of the large epiotics which meet in the middle line (Pl. X, fig. 1, *epo.*). These bones, which have a triangular posterior face, form about half the depth of the occiput and rest directly on the exoccipitals (*exo.*), which also meet in the middle line above the foramen magnum (*f.m.*). In the fossil shown in Pl. X, figs. 1, 1 *a*, the exoccipitals are slightly crushed above, so that the epiotics form an overhanging ledge, but the general shape of the bones is well indicated. Each exoccipital ends postero-inferiorly in an occipital condyle, and its concave posterior face slopes upwards and forwards, separated by a sharp angulation from its lateral face, which is extensive, since the bone enters for a considerable distance into the lateral wall of the brain-case, as in the salmon (Pl. X, fig. 1 a, exo.). A large oval foramen for the exit of the vagus nerve (x) is conspicuous in this lateral face. The lower limit of the exoccipital is not seen in the fossil represented in Pl. X, figs. 1, 1 a, owing to the water-worn condition of the basioccipital (bo.), of which only the upper part, forming the roof of the notochordal cavity, is preserved. Its upper limit is also obscure laterally, but it appears to be capped by a small opisthotic (opo.), which forms the floor of the diminutive temporal fossa, and unites inside with a lateral prominence of the epiotic. The exoccipital and opisthotic are suturally united in front with a large pro-otic element (pro.), which doubtless meets the epiotic postero-superiorly, and is capped by a small sphenotic (or post-

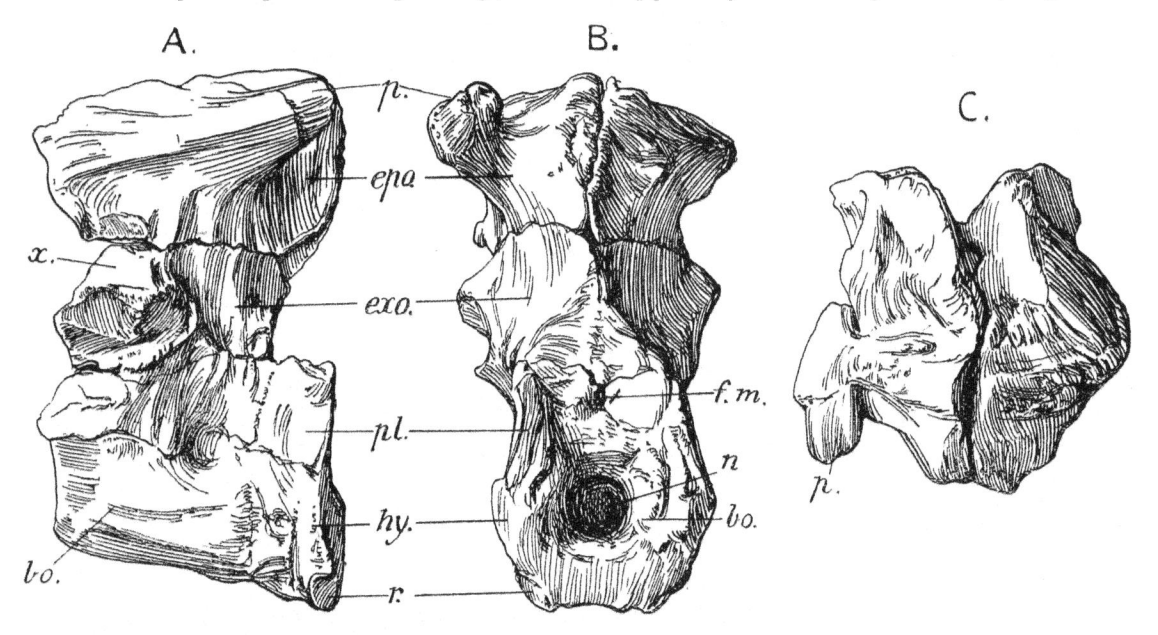

Fig. 15.—*Lepidotus* sp.; occipital portion of skull, left lateral (A), posterior (B), and superior (C) views, nat. size.—Oxford Clay: Peterborough. Leeds Collection (B. M. no. P. 9998). bo., basioccipital; epo., epiotic; exo., exoccipital; f.m., foramen magnum, reduced by distortion; hy., hypocentrum of first vertebra; n., pit in basioccipital for notochord; pl., pleurocentrum of first vertebra; p., process of epiotic, nature undetermined; r., facet on hypocentrum for articulation of rib; x., facet probably for opisthotic.

frontal) antero-superiorly. The sphenotic is not clearly seen in the original of Pl. X, fig. 1 a, but is partly shown in another specimen in the British Museum (no. P. 6342).

It is interesting to note that a similar arrangement of the epiotic bones in the occiput has been discovered by Mr. Alfred N. Leeds in a skull of *Lepidotus* from the Oxford Clay of Peterborough (Brit. Mus., no. P. 9998). The various elements are somewhat crushed and broken (Text-fig. 15), but the epiotics (epo.) are complete, and can be handled separately, proving that they meet in the middle line not only behind but above the brain-cavity throughout their entire length. They rest directly on the exoccipitals (exo.), which are fractured behind and so much crushed as almost to obscure the foramen magnum (f.m.). The exoccipital extends considerably into the lateral wall of the brain-case, and bears an articular surface antero-superiorly (*x.*) probably for a small opisthotic, which is missing in the fossil. Posteriorly it is fused with the pleurocentrum (*pl.*) of the first ring-vertebra. Inferiorly the exoccipital is firmly united with the large basioccipital (*bo.*), which is grooved below for the basicranial canal, pierced behind by a deep excavation for the notochord (*n.*), and fused with the large hypocentrum (*hy.*) of the first vertebra. This hypocentrum bears a pair of prominent facets (*r.*) for the ribs. The pro-otic bone of the same specimen is relatively large and must have articulated both with the exoccipital and the opisthotic, while it was capped in front with a sphenotic or postfrontal.

It may be added that the epiotics unite in a median suture above the exoccipitals in some existing Teleostean fishes, such as *Acanthurus*.

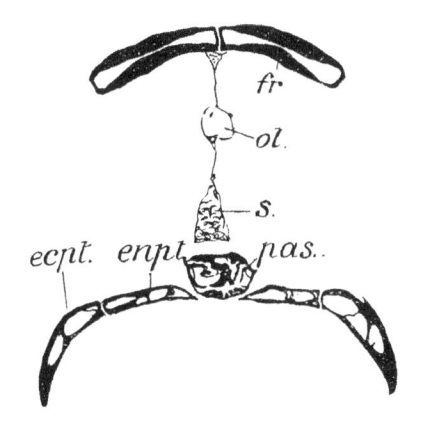

Fig. 16.—*Lepidotus mantelli*, Agassiz; transverse section of skull between orbits, two-thirds nat. size.—Wealden: Hastings. Beckles Collection (B. M. no. P. 6342). *ecpt*, ectopterygoid; *enpt.*, entopterygoid; *fr.*, frontal; *ol.*, interorbital passage for olfactory nerves; *pas.*, parasphenoid; *s.*, interorbital septum.

There are also ossifications in the interorbital septum, but their precise character is uncertain. A transverse section of one skull at the middle of the orbit is shown in Text-fig. 16. In this position the septum ends above and below in very coarsely cancellated bone, triangular in transverse section (*s.*), while the intervening part is merely a thin, ossified lamina, which widens in the middle into a tube, doubtless for the passage of the olfactory nerves (*ol.*). Below the interorbital septum the parasphenoid is seen (*pas.*), with the superficially ossified hinder part of the entopterygoids (*enpt.*) and ectopterygoids (*ecpt.*). Further forwards there are indications of bone in the ethmoid region in several specimens, with apparently two separate tubes for the passage of the olfactory nerves.

The external bones are thick and vary much in appearance according to their state of abrasion in the fossils. They are all more or less rugose, and unworn specimens exhibit a variable extent of tuberculation. The parietal bones (Pl. VIII, fig. 2 *a*; Pl. IX, figs. 1, 1 *a*; Pl. X, fig. 3; *pa.*) form an unsymmetrical pair, of which one is both wider and longer than the other; they unite in a nearly straight median suture, which has sometimes one small sinuosity behind; and their superficial rugosity is both coarsest and strongest near the outer border, especially behind, where the slime-canal traverses the bone in an L-shaped groove. The squamosal (*sq.*) on each side is a narrower bone, extending as far forwards as the parietals, but not quite so far backwards. It unites with the adjacent parietal in a slightly wavy suture, and its outer face is very coarsely rugose, while the groove for the transverse slime-canal behind is well marked. The squamosal is narrowest in front, where it bounds the hinder end of the frontal, and meets the circumorbital ring, from which a small dermo-postfrontal is sometimes detached. The frontals (Pl. VIII, fig. 2 *a*; Pl. IX, figs. 1, 1 *a*; Pl. X, fig. 3; *fr.*) are nearly three times as long as the parietals, and unite in a very slightly wavy median suture. Each is widest behind, the maximum width being about a third of the total length; and its comparatively narrow anterior portion ends in a few pointed digitations. The absence of these digitations in the frontal ascribed to *Lepidotus fittoni* by Agassiz (Poiss. Foss., vol. ii, pt. 1, p. 264, pl. xxx *b*, fig. 3) is due to the imperfection of the specimen. The outer margin of the bone immediately in front of the squamosal is slightly indented by the overlap of the three upper plates of the circumorbital ring. Its outer face is usually rugose and tuberculated only in the hinder half, where the markings are coarsest near the outer margin; it then tends to rise into a rounded boss at the median suture between the orbits; and the slender anterior half is nearly smooth, only with a longitudinal channeling. The nasals and other dermal bones of the ethmoid region are unknown.

The cheek is completely covered with plates, which form a circumorbital ring (Pl. VIII, figs. 2, 2 *a*; Pl. IX, fig. 1; Pl. X, fig. 3; *co.*) bounded behind by postorbitals (*po.*), and continued in front by a few preorbitals (*pro.*). They are all more or less rugose and tuberculated, but the most delicate and easily destroyed markings are on the upper postorbitals. Of the circumorbital ring, three plates occur above the eye, the hindmost being the largest and sometimes transversely subdivided, so that a separate piece in contact with both the squamosal and the frontal may be regarded as a dermo-postfrontal (*e. g.* B. M. no. P. 6338). The two posterior circumorbitals, though of irregular shape, are about as deep as wide, while the other four plates completing the ring antero-inferiorly are much deeper than wide. The slime-canal is not conspicuous. There are at least four plates in the preorbital series. The irregularly pentagonal upper postorbital (see especially Pl. IX, fig. 1, *po.*) is comparatively large, in contact with the squamosal and two circumorbitals; but the other plates of the postorbital series, very irregular in shape, usually from five to seven in number, though sometimes further subdivided, are not wider than the circumorbitals. The foremost and lowest plate is always longer than wide and tapers to a blunt point below the foremost circumorbital.

The mandibular suspensorium is inclined forwards so that the quadrate articulation is beneath the middle of the orbit. The hyomandibular (Pl. VIII, fig. 1, *hm.*) is a narrow lamina of bone, with a broad prominence behind for the support of the operculum, and strengthened below this on the outer face by a longitudinal ridge. The metapterygoid (Pl. VIII, fig. 1, *mpt.*) is imperfectly known, but is a relatively large thin lamina of bone probably shaped almost as in *Amia*. The entopterygoid has already been mentioned as a thin toothless lamina seen in section in Text-fig. 16, *enpt*. The ectopterygoid, though thin and toothless at its hinder end (Text-fig. 16, *ecpt.*) and rising above into a longitudinal sharp

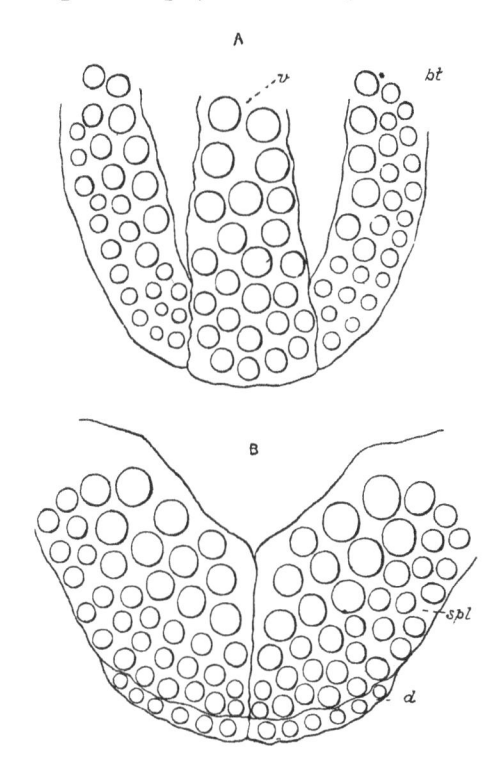

FIG. 17.—*Lepidotus mantelli*, Agassiz; diagram of arrangement of teeth in upper (A) and lower (B) jaws, about nat. size.—Wealden: Sussex. *d.*, dentary; *pt.*, pterygo-palatine; *spl.*, splenial; *v.*, vomer. From Catal. Foss. Fishes, Brit. Mus., pt. iii, 1895.

crest throughout its length (Pl. XI, fig. 3 *a*), is much thickened at the oral border to bear the powerful dentition, and seems to be completely fused in front with the equally thickened palatine, which abuts against the vomer. The slightly concave oral face of the pterygo-palatine thus formed bears the irregular longitudinal series of teeth, of which the inner are the largest and the most anterior are the smallest (Pl. XI, figs. 3, 4). An outer fourth row of very small teeth is sometimes present. The single vomer (Pl. XI, figs. 2, 4) is also much thickened (fig. 2 *b*), and its long and narrow oral face bears teeth almost as far back as the hinder limit of teeth on the ectopterygoids. The vomerine teeth are irregularly arranged, but the largest are in two pairs behind, and they decrease in size forwards, where they are in four or even five longitudinal rows. The maxilla is a relatively small and delicate lamina of bone, forming a deep expansion behind but tapering forwards

(Pl. IX, fig. 1, *mx.*). Its outer face is smooth, and teeth have not been observed on the oral border in the known specimens. A small narrow supramaxilla, pointed behind, occurs above its hinder expansion (*sma.*). The premaxillæ are known only in the small specimen shown in Pl. X, figs. 3, 3 *a*, *pmx.*, where they are shaped nearly as in *Lepidotus minor* (see p. 30). The broad anterior ascending process of each bone clearly passes up beneath the anterior end of the frontal; and the comparatively small portion extended along the oral border bears a row of styliform teeth, each capped by enamel. The mandible is short and stout, much deepened in the coronoid region, much horizontally expanded at the symphysis. As seen from the outer face (Pl. IX, fig. 1, *ag.*), the angular bone is especially short and deep, smooth above but more or less coarsely rugose below, above the longitudinal groove for the passage of the slime-canal. It is capped by a very small coronoid bone (B. M. no. P. 6342). The hinder ascending part of the dentary (*d.*) is as deep as the angular and coronoid bones together, and its outer convex face is either smooth or faintly rugose. Just in front of this ascending part the bone is of least depth, and it then deepens slightly again towards the symphysis (Pl. VIII, fig. 2 *b*), which slopes sharply backwards, sometimes approaching a horizontal plane for the support of the massive splenial. The oral border in front bears a single row of six to nine comparatively small but stout styliform teeth (Pl. XI, fig. 6); and the outer face of the bone is more or less coarsely rugose, with the course of the slime-canal marked by a row of large pits. The tooth-bearing end of a very small dentary has already been described under the name of *Tetragonolepis mastodonteus* by L. Agassiz, Poiss. Foss., vol. ii., pt. i (1837), p. 216, pl. xxiii *e*, figs. 3, 4. The massive splenial (Pl. XI, fig. 5, *spl.*) meets its fellow in an extensive symphysis, and its slightly concave oral face is covered by five or six rows of tritoral teeth, which are largest within. There is some ossification in the meckelian cartilage, but its shape and extent are uncertain.

All the teeth are fused with the supporting bone, not in sockets, and they have a large pulp-cavity, from which very minute, irregularly branching tubuli radiate into the dentine. Successional teeth are abundant in the thick cancellated bone beneath the functional teeth (Pl. XI, fig. 1), and they clearly turn through an angle of 180° in the course of development as in other species of *Lepidotus* (Pl. XI, figs. 1, 3 *b*, 5 *b*). The enamelled crown, when unworn, rises to a sharp median apex; and occasionally, as in the hinder teeth of Pl. XI, fig. 2, there is a slight apical constriction which tends to make it mammilliform (fig. 2 *c*). Even in the type specimen of *Lepidotus fittoni* (contrary to the statement by Agassiz) some of the inner teeth exhibit the median apex. When worn they become first rounded, then flattened, and eventually sometimes have the pulp-cavity exposed; but there is no uniformity in the degree of wear in any group of teeth, and they appear to be shed in indefinite order. In all undoubted specimens of *L. mantelli* the enamel of the dental crown is smooth; but in one example of upper dentition from the Isle of Wight (Pl. XI, fig. 4) some of the teeth exhibit faint irregular wrinkles radiating from the apex to an encircling wrinkle which forms a kind of cingulum round the base (figs. 4 a, 4 b).

In the hyoid arch (Pl. X, fig. 2), the epihyal (*eph.*) is comparatively large and stout, bearing about six branchiostegal rays, of which the uppermost (*br.*) forms a long and narrow delicate lamina. The ceratohyal (*ch.*) is less than twice as long as the epihyal, comparatively small in front of its constriction, and without branchiostegal rays. The hypohyal (*hyh.*) is a short and stout bone, also constricted in the middle.

The branchial arches are unknown, but delicate calcified gill-supports are seen in a specimen in the Beckles Collection (B. M. no. P. 6343).

The preoperculum, best seen in Pl. IX, fig. 1, *pop.*, is narrow and gently curved at the angle. It is much overlapped by the postorbital cheek-plates, but extends upwards to the hinder end of the squamosal in the cranial roof. Its ascending limb is especially narrow, and marked with a slight rugosity or tuberculation only near its upper end. Its lower limb is wider, with an irregular coarse rugosity radiating downwards and backwards from the deep groove for the slime-canal. The operculum and suboperculum vary considerably, but some of the differences observed are probably due to imperfections in preservation. As shown in Pl. IX, fig. 1, *op.*, the operculum is somewhat wider below than above, and its maximum width is about two-thirds of its depth. Its anterior margin is sigmoidally curved, and its outer face is finely tuberculated, especially in the upper half. In the type specimen of *Lepidotus mantelli* (Pl. VIII, fig. 1, *op.*), the operculum appears to be narrower, but this may be due at least in part to crushing and fracturing. In the type specimen of *L. fittoni* (Pl. VIII, fig. 2, *op.*), the bone is broken above and behind, obscured in front with ironstone, and so much abraded that its outer surface appears to be only rugose. In other abraded specimens its maximum width scarcely equals two-thirds of its depth. The suboperculum (Pl. VIII, fig. 2; Pl. IX, fig. 1; *sop.*) is usually more than half as deep as wide, with a relatively large anterior ascending process, and the outer face very feebly ornamented either with rugæ or tubercles. In the type specimen of *L. mantelli*, however, the suboperculum (Pl. VIII, fig. 1, *sop.*) is not half so deep as wide, and its outer face is strongly tuberculated. The interoperculum (Pl. IX, fig. 1, *iop.*), is much extended and tapering forwards, with the ornament as variable as that of the other bones. Not more than six branchiostegal rays have been clearly seen, and the uppermost is relatively large (Pl. X, fig. 2, *br.*). There is no gular plate.

The notochord must have been persistent, but at least in the anterior part of the abdominal region there are surrounding ossifications. These are only imperfectly known in a few weathered specimens, of which the best is shown in Text-fig. 18. Encircling the upper part of the space left vacant by the decay of the notochord and its sheath are two sectors of spongy bone (*pl.*) which thicken upwards, until they nearly surround the canal for the spinal cord and form the expanded base of the neural arch (*n.a.*). Two similar but larger sectors of spongy bone (*hy.*) encircle the lower part of the notochordal space, each bearing a short transverse process or parapophysis (*t.*) for the support of the rib. Each dorsal piece is directly continuous with the rod-shaped lamina of the neural arch, which is inclined backwards and does not appear to be fused with its fellow of the opposite side at the upper end, where it is in contact with the long rod-shaped neural spine (*n.s.*), also inclined backwards and curved so as to be a little concave anteriorly. The ribs are com-

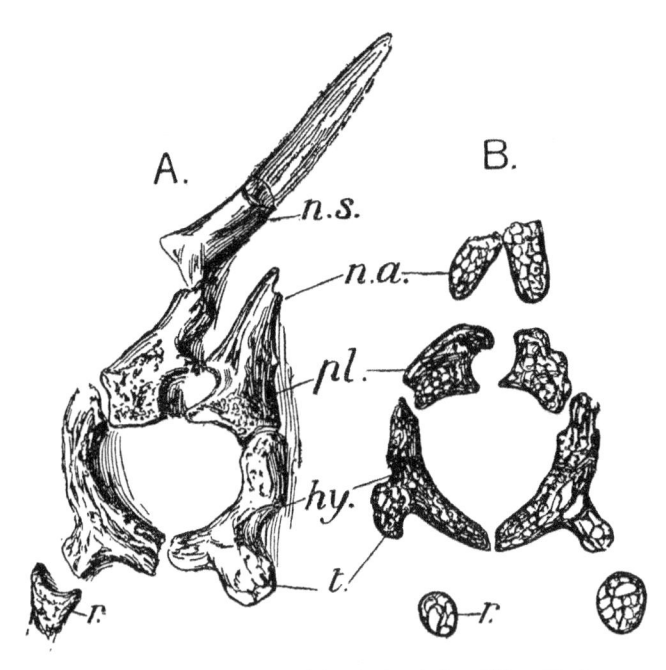

Fig. 18.—*Lepidotus mantelli*, Agassiz; anterior vertebra in front view (A) and in transverse section (B), nat. size.—Wealden: Hastings. Beckles Collection (B. M. no. P. 6348 c). *hy.*, hypocentrum; *n.a.*, neural arch; *n.s.*, neural spine; *pl.*, pleurocentrum; *r.*, rib; *t.*, process on hypocentrum for articulation of rib.

paratively slender and extend considerably more than half-way towards the ventral border of the fish.

The supratemporal plates are small, and usually in four pairs, as shown in the type specimen of *L. fittoni* (Pl. VIII, fig. 2 *a*, *st.*), and in the immature head represented in Pl. IX, *st*. Two, however, are sometimes fused together, so that the series becomes unsymmetrical (Pl. VIII, fig. 3). All are marked with a coarse and irregular rugose ornament, with occasional indications of the traversing slime-canal. Each of the inner pair of plates is wider than long, and extends along at least half of the parietal border. The next two plates are much smaller, being longer than wide, the inner plate in contact solely with the parietal, the outer plate bordering an angle between the parietal and squamosal bones. The outermost plate of the series, which is slightly larger than the two latter, bounds both the hinder border of the squamosal and the greater part of the upper border of the operculum. In the specimen shown in Pl. VIII, fig. 3, the two inner supratemporal plates are fused together on both sides, while the two outer plates are fused together only on the right side. The exposed face of the post-temporal (Pl. VIII, figs. 2 a, 3 ; Pl. IX, figs. 1, 1 a ; Pl. X, fig. 3 ; *ptt.*) is irregularly triangular in shape, wider than long, with its inner apex extending to the second supratemporal plate; it is also coarsely rugose. The supraclavicle is a deep and narrow plate of bone, fragmentary in the type specimen (Pl. VIII, fig. 1, *scl.*) but well shown in Pl. IX, fig. 1, *scl*. It is truncated and thickest at its upper end where it articulates with the post-temporal; while its outer face and hinder border are smooth, except in the upper part, where there is some coarse rugosity posteriorly. The clavicle is relatively large and curves well forwards, with its narrow exposed portion smooth, only slightly notched at the place of origin of the pectoral fin (Pl. VIII, fig. 1 ; Pl. IX, fig. 1 ; *cl.*). As seen in the type specimen, the anterior edge of the exposed portion is marked as usual by a few rows of small granulations; and as seen in a detached specimen in B. M. no. 23624, the thin inner or anterior wing of the bone is so wide that the maximum width of the clavicle equals one-third of its depth. The postclavicular plates are smooth, with no enamel except occasionally in the form of a slight tuberculation. The upper postclavicular, which is nearly complete in the type specimen (Pl. VIII, fig. 1, *pcl.*), is deep and narrow, and its tapering upper end bounds the lower part of the supraclavicle. In this specimen it is tuberculated, but in the original of Pl. IX, fig. 1, it is entirely smooth so far as preserved. The second plate, directly beneath the first, is best seen in the type specimen of *L. fittoni* (Pl. VIII, fig. 2, *pcl.*), where it is also slightly tuberculated; it is nearly as wide as deep, and irregularly triangular in shape with a truncated apex. The series is completed below by two relatively small and narrow plates, of which the lower is the larger and fits into the notch of the clavicle at the origin of the pectoral fin (as shown in Pl. IX, fig. 1, *pcl.*). At least part of the endoskeleton of the pectoral arch is well ossified, and a bone which may probably be identified as coracoid occurs in the type specimen (Pl. VIII, fig. 1, *cor.*). It is constricted postero-superiorly into a wide and thickened pedicle which ends in an articular facette. Remains of long and slender basal bones or radials are scattered near the pectoral fin in the same specimen, and at least five of these elements in series are also seen in another specimen in the British Museum (no. 23624). As shown in the type specimen, the rather slender pectoral fin-rays have a long unsegmented basal part (Pl. VIII, fig. 1, *pct.*); and the stout, deeply-overlapping fulcra on the anterior ray are in double series, while two relatively large basal fulcra are directly inserted in line with the fin-rays. The number of pectoral fin-rays is unknown. The pelvic fins are much smaller than the pectorals, and inserted nearer to the latter than to the anal fin. They are fringed with a paired row of small fulcra. The dorsal and anal fins are best shown in a fine specimen discovered by Mr. Charles

Dawson (Pl. VII, fig. 7), where they are equally elevated, their greatest height measuring about two-thirds of the depth of the trunk at the insertion of the dorsal. The enamelled biserial fulcra of the dorsal fin are comparatively large, and four or even five (B. M. no. P. 6336) of these fulcra, of gradually increasing length, are directly inserted in the ridge of the back. The dorsal fin-rays, about fourteen in number, decrease in length backwards, and each is very finely divided and articulated for more than half of its length. The anal fin is closely similar to the dorsal, but with smaller biserial fulcra, only about ten rays, and the undivided bases of these rays relatively short. The caudal fin is known only by its basal part, which resembles that in other species (Pl. VII, fig. 7).

All the scales are usually smooth, sometimes with a faintly concave face, but there is a tendency to coarse oblique grooving, especially when the enamel is partly removed and the subjacent structure which causes this grooving becomes evident (Pl. VIII, fig. 4). Irregularly tuberculated or rugose scales have only been observed in the anterior dorsal region of one small specimen (Pl. IX). The scales on the anterior part of the abdominal region, especially on the flank (Pl. XI, fig. 7), are usually finely serrated, but the degree of serration varies considerably from the extreme in the original of Pl. IX, through the apparently partial serration in the type specimen (Pl. VIII, fig. 1), to the more irregular serration in the so-called *L. fittoni* (Pl. VIII, fig. 2). In the scales of other parts the hinder margin is smooth. As counted along the lateral line, the total number of transverse series of scales is about forty, while the number in a series above the origin of the pelvic fin is nineteen or twenty. The principal flank-scales in the abdominal region are somewhat deeper than broad, and those of the lateral line, in the eleventh or twelfth row above the pelvic fins (Pl. IX, fig. 1, *l.*), are notched just above their postero-inferior angle and usually pierced by an oval foramen near their anterior margin. The scales more dorsally and ventrally, and in the caudal region, are nearly equilateral or even broader than deep; and many of those in the dorsal region between the occiput and the dorsal fin are pierced by a large foramen for the exit of slime-apparatus, without any definite arrangement in lines (Pl. VII, fig. 7; Pl. VIII, figs. 2 *a*, 3). Beneath the base of the dorsal and anal fins the scales become small and irregular, with rounded angles, and with enamel often failing to extend to the edge. At the base of the caudal fin the scales are also small and irregular, but rather elongate-rhomboidal in shape. The dorsal ridge-scales are not much, if at all, enlarged; but most of them are acuminate, and they vary much in shape from the slight prominence seen in Pl. VIII, fig. 3, to the extreme acumination made conspicuous by crushing in the original of Pl. VII, fig. 7. The ventral ridge-scales are neither enlarged nor more acuminate than the other scales; the usual enlarged scales occur near the origin of the anal fin.

All the scales are thick, and those of the abdominal region are united in the usual manner by deep overlap and a peg-and-socket articulation (Pl. XI, figs. 7, 8).

This articulation, however, is comparatively feeble, while the inner vertical ridge connected with it is low and irregularly widened (fig. 8). On the middle of the flank, the upper process of the wide overlapped margin of each scale is longer and more slender than the lower process (figs. 7, 8). In the dorsal, ventral, and caudal scales, the inner face is tumid, and there is no peg-and-socket articulation (Pl. XI, figs. 10 a—13 a). Some of the scales in the dorsal part of the abdominal region are nearly square, and their straight overlapped border is prolonged upwards into a strong process (Pl. XI, figs. 9, 10). In the more elongated scales of the ventral part of the abdominal region the overlapped margin is forked as in the scales of the flank, with the upper process much longer than the lower process and often inclined upwards (Pl. XI, figs. 11, 12). In the caudal region (Pl. XI, fig. 13), the anterior overlapped margin is comparatively narrow, and the inner face (fig. 13 a) remarkably tumid. An isolated scale of the lateral line in the caudal region (figs. 14, 14 a) shows especially well the two orifices on the outer face for the slime-canal and its posterior exit on the inner face.

Affinities.—*Lepidotus mantelli* appears to be most closely related to *Lepidotus lævis*, Agassiz, as interpreted by F. Priem, who describes the greater part of a fish from the Lower Kimmeridgian of Cerin (Ain), France, in Annales de Paléontologie, vol. iii (1908), p. 2, pl. i.

Horizon and Localities.—Wealden: Sussex and (?) Isle of Wight. Upper Purbeck Beds: Sussex.

Undetermined species of *Lepidotus* about as large as *L. mantelli* are known by fragments from the Middle Purbeck Beds of Swanage, Dorset. Part of a trunk about 25 cm. in maximum depth in the Dorset County Museum, has smooth scales without serrations. The opercular and anterior region of an equally large fish in the British Museum exhibits a coarse irregular serration of the smooth scales.

One or more undetermined dwarf species of *Lepidotus* also occur in the Purbeck Beds of the Vale of Wardour, Wiltshire.

Family PYCNODONTIDÆ.

Genus **ATHRODON**, Sauvage.

Athrodon, H. E. Sauvage, Bull. Soc. Géol. France [3], vol. viii, 1880, p. 530.

Generic Characters.—Splenial bone unusually stout, with a deep symphysial facette; its oral face covered with rounded teeth, which are arranged in more or less irregular longitudinal series, the one principal series generally not well differentiated from the others.

Type Species.—*Athrodon douvillei* (H. E. Sauvage, Bull. Soc. Géol. France [3], vol. viii, 1880, p. 530, pl. xix, fig. 5) from the Lower Portlandian of Boulogne, France.

Remarks.—This genus is still known only by the dentition, which appears to be less specialised than that of any other Pycnodont. It ranges from the Kimmeridgian to the Senonian in Western Europe.

1. **Athrodon intermedius**, A. S. Woodward. Text-figure 19.

1893. *Athrodon intermedius*, A. S. Woodward, Geol. Mag. [3], vol. x, p. 434, pl. xvi, fig. 1.

Type.—Left splenial dentition; British Museum.

Specific Characters.—Splenial bone comparatively elongated, with closely arranged teeth, mostly smooth and nearly round, a few exhibiting an apical pit with feebly crimped margin, disposed in about five or six irregular longitudinal series, the largest forming a principal series near the symphysial margin of the bone.

Description of Specimen.—The only known example of this species is shown of the natural size in Text-fig. 19. Many of the teeth have been much worn during life, but those of the largest series seem to have been gently rounded, not pitted. Some of the marginal teeth are broken away, but they appear to have been all small. The slight apical pit is well seen in the hindmost tooth of one of these series.

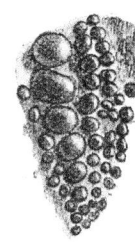

FIG. 19.—*Athrodon intermedius*, A. S. Woodward; left splenial dentition, oral aspect, nat. size. — Purbeck Beds: Aylesbury. The type specimen (B. M. no. 40314).

Horizon and Locality.—Purbeck Beds: Aylesbury, Buckinghamshire.

Genus MESODON, Wagner.

Mesodon, A. Wagner, Abhandl. k. bay. Akad. Wiss., math.-phys. Cl., vol. vi, 1851, p. 56.
(?) *Typodus*, F. A. Quenstedt, Der Jura, 1858, p. 781.
Macromesodon, J. F. Blake, Mon. Fauna Cornbrash (Pal. Soc., 1905), p. 32.

Generic Characters.—Trunk discoidal, not produced at the caudal pedicle. Head and opercular bones more or less ornamented with granulations; cleft of mouth very oblique; teeth smooth, or with feeble indentation and rugæ; vomerine teeth in five longitudinal series, the lateral pairs often irregular; splenial bone with symphysial facette not deepened; splenial dentition comprising one principal series of teeth with three or more outer series and one or two inner series, usually irregularly arranged. Neural and hæmal arches of axial skeleton not sufficiently expanded to meet round the notochord. Fin-rays robust, articulated, and divided distally. Pelvic fins present; dorsal and anal fins deep throughout their extent, the former occupying the hinder half of the back, and the latter somewhat shorter; caudal fin fan-shaped, with a truncated or convex hinder margin, and arising immediately between the posterior extremities of the dorsal and anal fins. Scales usually smooth, covering only the anterior half of the trunk, and complete only in the lower part of the flank.

Type Species.—*Mesodon macropterus* (A. Wagner, *loc. cit.*, 1851, pp. 49, 56,

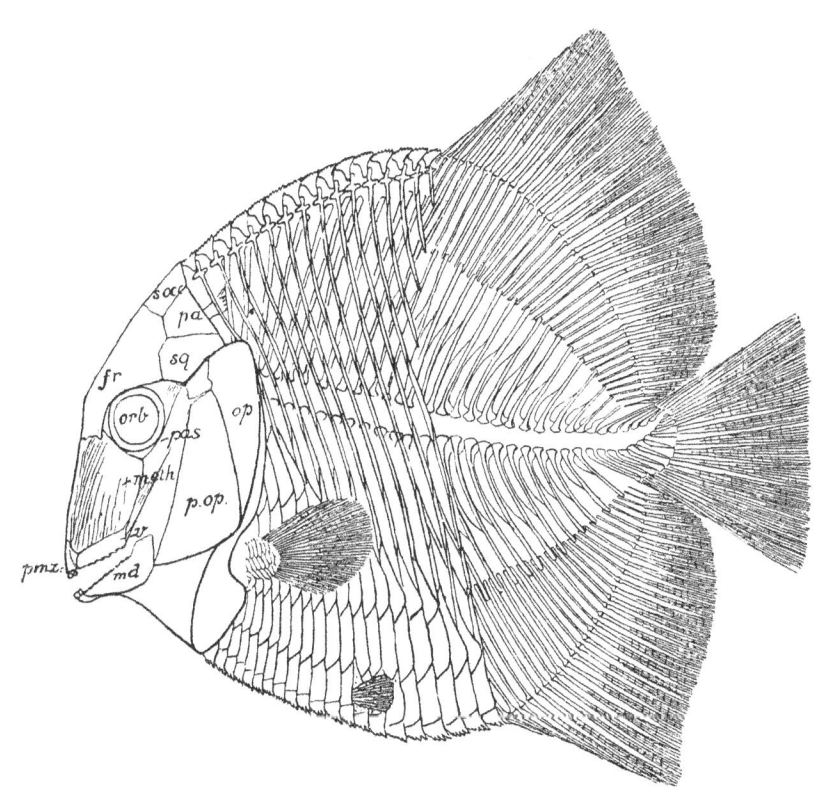

Fig. 20.—*Mesodon macropterus* (Agassiz); restoration, about two-thirds nat. size—Lower Kimmeridgian (Lithographic Stone); Bavaria. *fr.*, frontal; *m.eth.*, mesethmoid; *md.*, mandible, showing narrow dentary in front; *op.*, operculum; *orb.*, orbit; *p.op.*, preoperculum; *pa.*, parietal; *pas.*, parasphenoid; *pmx.*, premaxilla; *s.occ*, supraoccipital; *sq.*, squamosal; *v.*, vomer.

pl. iv, fig. 2) from the Lower Kimmeridgian (Lithographic Stone) of Kelheim, Bavaria. The species had previously been named *Gyrodus macropterus* by L. Agassiz, Poiss. Foss., Feuill. (1834), p. 18, and vol. ii, pt. ii (1844), p. 301.

Remarks.—In a restoration of the type species of *Mesodon* published in my 'Outlines of Vertebrate Palæontology' (1898), the arrangement of the scales is wrongly based on the Liassic species now named *Eomesodon liassicus* (see below, p. 55). A new restoration is accordingly given in the accompanying Text-fig. 20.

In a restoration of *Mesodon bernissartensis*, from the Wealden of Belgium, published by R. H. Traquair in his "Poissons Wealdiens de Bernissart" (Ann.

Mus. Roy. Hist. Nat. Belg., vol. v, 1911, fig. 11, p. 30) the limits of the vomer and mesethmoid are erroneously indicated, the posterior digitate process of the parietal is omitted, and the scales at the origin of the pelvic fins are incorrectly shown. Crushed and broken specimens cannot be readily interpreted.

The geological range of typical species seems to be from the Kimmeridgian to the Wealden inclusive.

1. **Mesodon daviesi,** A. S. Woodward. Plate XII, figs. 1, 2.

1890. *Mesodon daviesi*, A. S. Woodward, Proc. Zool. Soc., p. 351, pl. xxviii, fig. 5.
1895. *Mesodon daviesi*, A. S. Woodward, Catal. Foss. Fishes B. M., pt. iii, p. 201.

Type.—Nearly complete fish; British Museum.

Specific Characters.—A species attaining a length of 25 cm. Maximum depth of trunk somewhat less than the length of the fish without caudal fin; head with opercular apparatus contained about three-and-a-half times in the same length; back gently rounded, and dorsal fin arising at the highest point. Principal splenial teeth rounded and smooth, about twice as broad as long, flanked externally by two series of smaller teeth, which are also smooth and often broader than long. Dorsal and anal fins about equally elevated, the latter with 29 or 30 supports and four-fifths as long as the former, which has 38 or 39 supports. Dorsal and ventral ridge-scales each with a row of four or five small denticles which increase slightly in size backwards.

Description of Specimens.—The type specimen (Pl. XII, fig. 1), with its incomplete counterpart, exhibits all the principal characters of the species, except those of the paired fins and the serration of the ridge-scales. A second smaller specimen, apparently of the same species, in the British Museum (no. P. 4381), has a slightly larger head, and agrees well with a more imperfect small specimen in the Dorset County Museum.

The head-bones as preserved in the fossils exhibit a fibrous texture, and the only external ornament is a radiating reticulation, without any tubercles. The parietal in the type specimen bears the usual large posteriorly-directed process with digitate extremity, and the supraoccipital ends abruptly, without any upward production. The orbit is as small as in the type species, and all the characteristic Pycnodont features are vaguely seen in the facial region. The small cleft of the mouth is, as usual, inclined slightly upwards and backwards. The vomerine dentition is seen only in edge-view, but in B. M. no. P. 4381 it is shown to have been slightly convex from side to side. The splenial dentition, partly exposed from the attached face in the same specimen, exhibits the principal teeth flanked outside by two series of teeth, which are also broader than long, while the inner slightly exceed the outer in size. As shown in the type specimen (Pl. XII, fig. 1 a), the transversely elongated principal splenial teeth are smooth and rounded, without any apical pit; the teeth of the first outer flanking series are also smooth, broader than long, and about half as broad as the principal teeth; while (as seen in the counterpart of the same specimen) the teeth of the second flanking series are somewhat smaller, but still broader than long. One tooth of the first flanking series exhibits an apical pit. The dentary bone (Pl. XII, fig. 2, *d.*), of the usual Pycnodont shape, bears two chisel-shaped teeth, of which the inner is the larger.

The triangular preoperculum (Pl. XII, figs. 1, 2, *pop.*) does not extend so far upwards as the relatively small, deep, and narrow operculum (*op.*); and when well preserved the radiating reticulated ornament of its outer face is very conspicuous (fig. 2, *pop.*). Traces of calcified gill-supports are seen beneath it in the Dorset Museum specimen.

The vacant space for the notochord forms a nearly straight band somewhat above the middle of the trunk, and the vertebral arches above and below it exhibit a small basal triangular expansion. In the type specimen (Pl. XII, fig. 1) 33 or 34 neural arches can be counted, the foremost being the stoutest and most widely spaced, while the hindmost three or four are comparatively diminutive and related to the support of the caudal fin. Most of these arches bear traces of the anterior laminar expansion (seen especially in the counterpart of the type), and about 18 may be reckoned as belonging to the abdominal region. The ribs are much expanded in their proximal portion and stout, but do not reach the ventral border of the trunk. There are about 13 hæmal arches in advance of the tail in the caudal region, nearly symmetrical with the opposed neurals; and at least 7 comparatively short hæmals are crowded together within the base of the caudal fin.

Of the paired fins, only traces of the pectoral are seen in the type specimen. Seven stout hour-glass-shaped supports are preserved in the basal lobe, and the expanse of the fin must have been large, with numerous closely articulated rays. Thirty-nine supports can be counted in the dorsal fin, and about 30 supports in the anal fin; and the dorsal fin is sufficiently well preserved to show that the foremost three or four rays gradually increase in length to the longest. In both these fins the articulations of the rays are close, and the unarticulated base is very short. In the caudal fin there are slightly more than 20 rays; of which those at the upper and lower borders are much crowded, and may be described as partly fulcral.

Remains of the scales cover the whole of the anterior half of the trunk to the origin of the median fins, arranged in 14 transverse series, with traces of the upper end of two more series just in front of the dorsal fin. For the greater part of their length the series of scales are represented in the fossils merely by the thickened inner rib which bears the peg-and-socket articulation. It is only in the lower part of the abdominal region that they are complete. Here they are thick and smooth, with an especially wide inner rib and large peg-and-socket articulation. The ventral ridge-scales are saddle-shaped and (as shown in the Dorset Museum specimen) each bears a row of four or five small denticles which increase slightly in size backwards. The lowest flank-scales are scarcely deeper than wide, though very irregular in shape. Above these the flank-scales are much deeper than wide, complete to the number of about four in the foremost series, but gradually reduced to one behind. All seem to be more or less irregular, as already described by Hennig in *Mesodon macropterus*.[1] So far as can be determined from the riblets, all the flank-scales are deep and must have been few in a transverse series. The course of the lateral line is not clear, but the upper slime-canal from the occiput to the origin of the dorsal fin is marked by slight transverse expansions on the successive riblets of deep scales immediately below the dorsal ridge. As shown in the Dorset Museum specimen, each dorsal ridge-scale is armed with a row of five small smooth denticles which increase slightly in size backwards. In the type specimen an irregularly elongate-triangular smooth scale, with its short anterior base crimped into three or four digitations (Pl. XII, fig. 1 d), occurs within the caudal fin and probably represents the last remnant of the squamation of the upper caudal lobe.

Horizon and Locality.—Middle Purbeck Beds: Swanage, Dorset.

2. Mesodon parvus, A. S. Woodward. Plate XII, figs. 3, 4.

1895. *Mesodon macropterus*, var. *parvus*, A. S. Woodward, Geol. Mag. [4], vol. ii, p. 147, pl. vii, fig. 2.

Type.—Fish with incomplete head; British Museum.

Specific Characters.—A small species, probably attaining a length of about 14 cm. Maximum depth of trunk somewhat less than the length of the fish without caudal fin; head with opercular apparatus contained about three times in the same length; back gently rounded, and dorsal fin arising slightly behind the highest point. Teeth frequently indented with a shallow pit; principal splenial teeth rounded and smooth, about twice as broad as long, flanked externally by two series of smaller teeth, the inner broader than long, the outer nearly round. Dorsal and anal fins equally elevated, the latter with about 26 supports and two-thirds as long as the former, which has about 36 supports. Dorsal and ventral ridge-scales with few comparatively large denticles, those of each scale rapidly increasing in size backwards.

Description of Specimens.—The type specimen (Pl. XII, fig. 3) is a fish scarcely more than 5 cm. in length; and a second specimen presented to the British Museum by Mr. T. T. Gething, showing the nearly complete head, is of about the same size. Portions of larger fishes, however, from the same formation and locality, probably belong to this species; one presented to the Museum of Practical

---

[1] E. Hennig, "*Gyrodus* und die Organisation der Pyknodonten," Palæontographica, vol. liii (1906), p. 172, fig. 8.

Geology (no. 3414) by the Rev. W. R. Andrews, representing an individual at least 14 cm. in length.

The head in Mr. Gething's fossil (Pl. XII, fig. 4) is well preserved in direct side view, with a rather large orbit, partly surrounded by the remains of an ossified sclerotic. The cranial roof is shown only in internal impression, but the supra-occipital is clearly not turned upwards behind, and the parietal bears the usual posterior digitate process. The facial region exhibits the mesethmoidal plate and the edge of the tooth-bearing vomer. The mandibular suspensorium is obscure both in this and in the type specimen, but a fortunate fracture displays the left splenial dentition from its attached face (fig. 4 a). The teeth preserved are in three regular series, the large principal teeth being about twice as broad as long, those of the next series also broader than long, but those of the outer series nearly round. There are traces of a shallow apical pit in the crown of two principal splenial teeth in the large specimen; and some of the lateral vomerine teeth are both pitted and faintly crimped. The two dentary teeth (fig. 4 a, d.), as usual, are chisel-shaped, and the outer of the two is comparatively small.

The maximum width of the nearly triangular preoperculum (Pl. XII, fig. 3, *pop.*) somewhat exceeds half its depth. The deep and narrow operculum (*op.*) is comparatively small. Both these bones are ornamented with sparse and partly reticulating ridges, which radiate backwards from a point on the front margin. Beneath the preoperculum are two branchiostegal rays (*br.*), of which the upper is the larger though both are relatively small. There are distinct traces of calcified gill-supports.

The axial skeleton of the trunk closely resembles that already described in *M. daviesi* (p. 51), but there seem to be only 11 hæmal arches in the caudal region in advance of the tail. The stout ribs are hollow in the fossils, and the two narrow wings in the upper half of each are distinct.

The pectoral fin, with the seven hour-glass-shaped supports in its basal lobe, is conspicuous on the flank above the lower expansion of the clavicle (Pl. XII, fig. 3). The comparatively small pelvic fin, with about 5 slender and much bifurcated rays, is also well seen in the type specimen (Pl. XII, fig. 3), inserted much nearer to the anal than to the pectoral fin. The dorsal and anal fins exhibit 36 and 26 supports respectively in both the small specimens, and their rays are remarkably slender and well-spaced, with less distal bifurcation than usual. The rays in the anal fin of the larger specimen, however, are stouter and more extensively bifurcated. Perhaps the first condition is a mark of immaturity. The caudal fin comprises 18 rays, with two or three slender fulcral rays at the origin above and below.

The scales are as in *M. daviesi* (p. 51), except that the smooth denticles on the dorsal and ventral ridges are coarser. Each dorsal ridge-scale bears three or four denticles which increase slightly in size backwards. Each ventral ridge-scale is more strongly armoured with only two or three denticles, of which the hinder is much the larger. The smooth complete scales on the lower portion of the flank, becoming reduced backwards, are especially well seen in the type specimen (Pl. XII, fig. 3), and their stout inner rib is still clearer in the second specimen (fig. 4). The dagger-shaped scales of the upper slime-canal are also distinct. The course of the lateral line is obscure in the abdominal region, but it is marked by a row of short tubular calcifications on the tail (fig. 3, *l.l.*). A small smooth rhombic scale occurs on the upper extremity of the caudal lobe.

Horizon and Locality.—Middle Purbeck Beds: Teffont, Wiltshire.

Genus **EOMESODON**, novum.

Generic Characters.—Profile of head especially steep and abdominal region of trunk much deepened; the caudal region relatively small. Head and opercular bones more or less coarsely granulated; jaws and teeth as in *Mesodon*, but with not less than three outer series of splenial teeth. Fins as in *Mesodon*. Scales complete over the whole of the trunk in advance of the median fins, not much deepened; ornamented with more or less coarse granulations.

Type Species.—*Eomesodon liassicus* (*Pycnodus liassicus*, Egerton, Figs. and Descript. Brit. Organic Remains, dec. viii—Mem. Geol. Surv., 1855—no. 10) from the Lower Lias probably of Barrow-on-Soar, Leicestershire, and other English localities.

Remarks.—The species referable to this genus have hitherto been included in *Mesodon*, but they form a group which is well distinguished by the deepening of the large abdominal region and the completeness of the abdominal squamation. The earliest species is *Eomesodon hoeferi*,[1] from the Upper Trias of Hallein, Salzburg, Austria. Next is the type species from the Lower Lias, and then follow the other Jurassic species, *Eomesodon rugulosus*,[2] *E. granulatus*,[3] *E. gibbosus*,[4] and *E. barnesi*, besides an uncertain number of species which are known only by the dentition.

The dorsal elevation of the anterior part of the trunk in the type specimen of *Eomesodon liassicus* is not well shown in Egerton's original figure of this fossil. It is therefore drawn again in the accompanying Text-fig. 21, which indicates some of the principal features of the genus and species. The skull is evidently that of a

---

[1] *Mesodon hoeferi*, D. G. Kramberger, Beitr. Paläont. und Geol. Oesterr. Ungarns, vol. xviii (1905), p. 219, pl. xx, fig. 5; pl. xxi, fig. 2.

[2] *Pycnodus rugulosus*. L. Agassiz, Poiss. Foss., vol. ii, pt. ii (1839—44), p. 194, pl. lxxii *a*, fig. 23; *Mesodon rugulosus*, A. S. Woodward, Proc. Geol. Assoc., vol. xii (1892), p. 239, pl. iv, figs. 2—4.

[3] *Pycnodus granulatus*, Graf zu Münster, Beitr. Petrefakt., pt. vii (1846), p. 44, pl. iii, figs. 11, 12; *Mesodon granulatus*, K. Fricke, Palæontogr., vol. xxii (1875), p. 359, pl. xviii, pl. xix, figs. 1—5.

[4] *Mesodon gibbosus*, A. Wagner, Abhandl. k. bay. Akad. Wiss., math. phys. Cl., vol. vi (1851), pp. 52, 56, pl. iii, fig. 2; E. Hennig, Centralbl. f. Min., 1907, p. 366, figs. 4, 5.

typical Pycnodont, with relatively large frontal bones, a small median supra-occipital plate (*socc.*), a quadrangular parietal (*pa.*), and a rather large squamosal (*sq.*), all closely ornamented with rows of granulations. One of the principal vomerine teeth, as already noted by Egerton, is coarsely crimped or tuberculated round the apex of the crown; a larger ovoid splenial tooth is smooth and not indented. The dentary bone bears two chisel-shaped teeth, of which the inner is the larger. Beneath the large triangular preoperculum there are two branchiostegal rays. In the axial skeleton of the trunk, the stout neural spines of the abdominal region do not reach the dorsal border; while neither neurals nor hæmals in the caudal region bear any laminar expansion. In the caudal region,

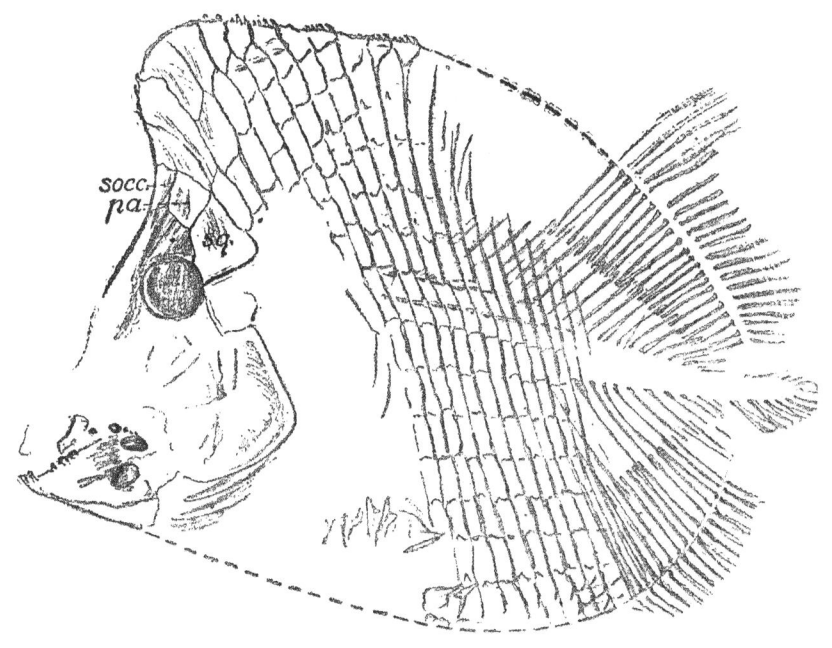

Fig. 21.—*Eomesodon liassicus* (Egerton); drawing of type specimen, nat. size.—Lower Lias; probably Barrow-on-Soar, Leicestershire. B. M. no. 19864. *pa.*, parietal; *socc.*, supraoccipital; *sq.*, squamosal.

however, as especially well seen in a smaller specimen (B. M. no. P. 1336, with counterpart in the Worcester Museum), each neural spine beneath the dorsal fin is double, a short straight rod arising from the front of the neural arch, the long neural spine proper arising behind and curving sharply backwards. The number of rays in the dorsal and anal fins seems to have been smaller than in the typical *Mesodon*. The large dorsal ridge-scales in the anterior part of the eminence are much deepened and modified, but they bear the usual median row of small hooked denticles, of which those in front are inclined backwards, while those behind the eminence seem to be upright or inclined forwards. The ventral ridge-scales are not seen in the type specimen; but three or four immediately in front of the anal fin in the smaller specimen already mentioned appear to be merely pointed and imbricating, without denticles. The lateral line is well marked by a ridge, and the course of the upper slime-canal is also distinct. In the deepest part of the trunk there appear to be about 8 scales in a transverse series above that traversed by the lateral line, and about 10 scales below this.

1. **Eomesodon barnesi,** A. S. Woodward. Plate XIII, fig. 1; Text-fig. 22.

1906. *Mesodon barnesi*, A. S. Woodward, Proc. Dorset Nat. Hist. Field Club, vol. xxvii, p. 187, pl. B, figs. 1—4.

Type.—Nearly complete fish; collection of the late F. J. Barnes, Esq.
Specific Characters.—Elevation of back sharply rounded, and maximum depth

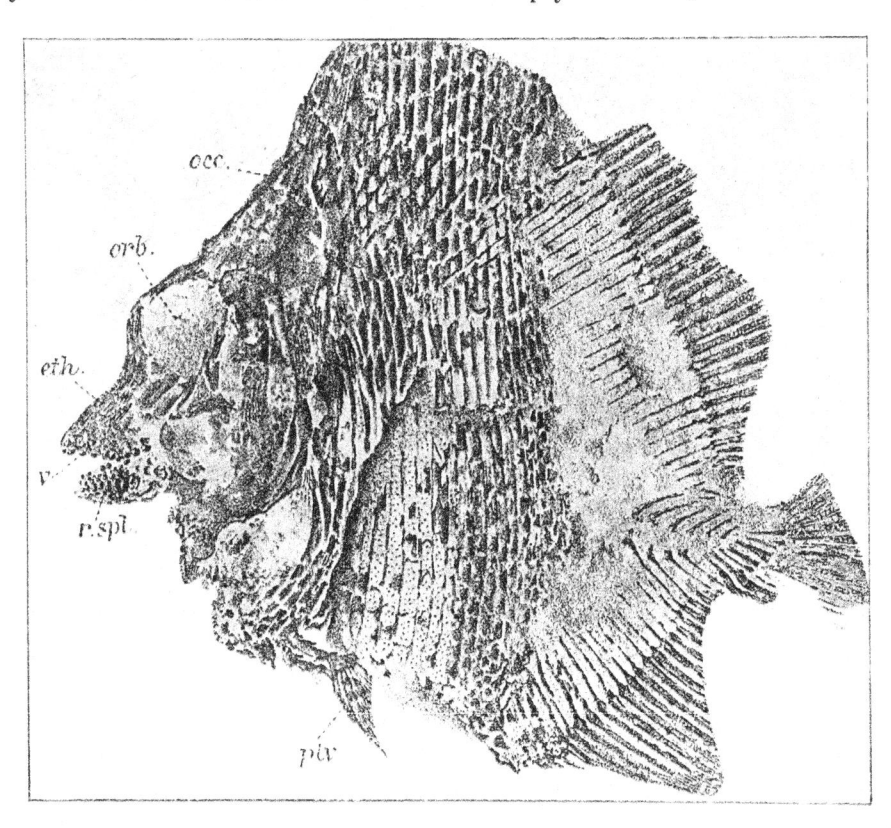

Fig. 22.—*Eomesodon barnesi*, A. S. Woodward; incomplete fish, two-thirds nat. size.—Portland Stone (Roach Bed); Portland, Dorset. *eth.*, mesethmoid; *occ.*, supraoccipital; *orb.*, orbit; *plv.*, pelvic fin; *r.spl.*, right splenial; *v.*, vomer. F. J. Barnes Collection.

of trunk about equalling the total length of the fish to the base of the caudal fin. Teeth smooth, a few of those of the lateral series having a faint apical pit with traces of crimping on the border; splenial teeth closely arranged, those of the principal series broader than long, flanked within by one row of small round teeth and externally by three series, of which the median is the smallest, and the outer about equal in size to the inner series. Dorsal fin with about 30, anal fin with about 20 rays. Granulation of external bones and scales coarse; enamelled denticles few and conspicuous on dorsal and ventral ridge-scales.

Description of Purbeckian Specimen.—The type specimen (Text-fig. 22) was obtained by the late Mr. F. J. Barnes, F.G.S., from the roach bed of the Portland Stone at Portland. The occurrence of the species in the Purbeck Beds is uncertain, but an imperfect specimen in the British Museum seems to agree with it in the parts which are comparable. This fossil (Pl. XIII, fig. 1), which is not much more than half as large as the type, is of approximately similar proportions, but lacks entirely the upper part of the head, the paired fins, and the caudal fin. The external bones and scales are ornamented with very prominent large hollow tubercles. The orbit is relatively large, and the cleft of the mouth small; and some of the lateral vomerine teeth are crimped round the shallow apical pit. As shown in broken section the bones are of very open texture. The typical rather stout neural and hæmal arches of the axial skeleton, with traces of their laminar expansions, are seen in the caudal region. The fragmentary dorsal and anal fins have clearly about 30 and 20 supports respectively, while all the rays are very stout, closely articulated, and subdivided distally. Some of the anterior rays of the anal fin are armed with rows of very small and slender, slightly arched denticles (fig. 1 c, a.). The squamation on the large and deep abdominal region is shown chiefly in impression, but a few actual remains of scales above the position of the notochord bear traces of the canal for the lateral line. The number of complete transverse series of scales cannot have been less than 20; and the scales, being less deepened, seem to have been more numerous in each series than in the typical *Mesodon*. Fragments of the dorsal ridge-scales bear very stout large smooth denticles (figs. 1 a, 1 b), and the ventral ridge-scales have these denticles still larger, each of the scales just in front of the anal fin bearing only one denticle with a boss in front (fig. 1 c).

Horizons and Localities.—Portland Stone (Roach Bed): Isle of Portland. (?) Middle Purbeck Beds: Swanage, Dorset.

2. **Eomesodon depressus,** sp. nov. Plate XIII, fig. 2.

Type.—Fragment of head and trunk; British Museum.

Specific Characters.—Elevation of back steep in front, but gradually descending to the rather remote dorsal fin; most of the dorsal ridge-scales bearing eight small denticles which are nearly uniform in size.

Description of Specimen.—Though well distinguished by the characters of the dorsal ridge, this species is known only by the fragmentary type specimen (Pl. XIII, fig. 2) which was discovered by the late Mr. Frederick Hovenden, F.G.S. The orbit (*orb.*) and the postorbital part of the skull are vaguely indicated, and the steep anterior profile of the fish does not appear to have been distorted. The very coarse tuberculation is visible on the fragments of head-bones preserved.

The axial skeleton of the trunk is nearly complete, comprising slightly more than 30 arches, of which about a third are caudal. Traces of the laminar expansions are seen both on the neural and the hæmal spines. A few of the dorsal fin-supports (*d.*) occur just behind the abdominal squamation, but no other remains of the fins are preserved. The scales are in slightly more than 20 regular transverse series, and exhibit both their internal and their external characters. The peg-and-socket articulation is very deep on all the scales, but the internal riblet is largest on the scales of the lower part of the abdominal region (fig. 2 *c*). Each scale is deeply overlapped in front, and its exposed portion is covered with large tubercles of ganoine (fig. 2 *b*). Each dorsal ridge-scale (fig. 2 *a*) is also provided along the middle line with a row of eight smooth conical denticles, which are nearly equal in size. The next row of scales below the dorsal ridge is traversed as usual by a slime-canal, and the lateral line is traceable along the middle of the flank. The number of scales in each transverse series is uncertain, but there seems to be about nine above the lateral line.

Horizon and Locality.—Middle Purbeck Beds: Swanage, Dorset.

Addendum.—Detached jaws of *Mesodon* and *Eomesodon* are not uncommon in the Purbeck Beds, but it is still not possible to name them specifically. The left splenial dentition shown in Pl. XIII, fig. 3, is remarkable for the crowding of its principal teeth and the irregularity of its inner and outer small teeth. Another left splenial dentition (Pl. XIII, fig. 4) may probably be referred to *Mesodon* rather than to *Microdon* on account of its well-developed inner row of small teeth, though this feature is not absolutely distinctive.

Genus **MICRODON**, Agassiz.

Microdon, L. Agassiz, Poiss. Foss., vol. ii, pt. i, 1833, p, 16, and pt. ii, 1844, p. 204.

Generic Characters.—Trunk deeply fusiform or discoidal, with short slender caudal pedicle. Cranial shield without supratemporal vacuities. Head and opercular bones ornamented with reticulating rugæ and pittings; two chisel-shaped teeth in each premaxilla and dentary; tritoral teeth smooth, sometimes feebly indented in the lateral series; vomerine teeth in five longitudinal series, but the inner lateral pairs regularly alternating with the widely spaced median teeth; splenial teeth in four series, the innermost being relatively small, the second the largest or principal series. Neural and hæmal arches of axial skeleton of trunk not expanding sufficiently to encircle the notochord. Fin-rays robust, articulated, and much divided distally. Pelvic fins present; dorsal and anal fins high and acuminate in front, rapidly becoming low and fringe-like behind, the former occupying at least the hinder half of the back, and the latter somewhat shorter, arising more posteriorly; caudal fin forked. Scales ornamented with reticulated rugæ or pittings, covering only the anterior half of the trunk, and complete only in the lower part of the flank; traces of riblets of scales sometimes on the middle of the flank of the caudal region.

Type Species.—*Microdon elegans* (L. Agassiz, Poiss. Foss., vol. ii, pt. i, 1833, p. 16, and pt. ii, 1839–44, p. 205, pl. lxix *b*) from the Lower Kimmeridgian (Lithographic Stone) of Solenhofen, Bavaria.

1. **Microdon radiatus**, Agassiz.   Plate XIV; Plate XV, figs. 1—5; Text-figure 23.

1839–44. *Microdon radiatus*, L. Agassiz, Poiss. Foss., vol. ii, pt. ii, p. 208, pl. lxix *c*, figs. 1, 2.
1840. *Microdon radiatus*. R. Owen, Odontography, p. 73, pl. xliii, fig. 1 [microscopical structure of teeth].
1895. *Microdon radiatus*, A. S. Woodward, Catal. Foss. Fishes Brit. Mus., pt. iii, p. 223.

Type.—Imperfect fish.

Specific Characters.—A species attaining a length of about 12 cm. Maximum depth of trunk somewhat less than total length to base of caudal fin; head with opercular apparatus occupying scarcely a quarter of total length of fish. Splenial teeth of principal series with well-rounded ends, wider than the two outer series, of which the outermost is considerably the larger. Vertebral axis at origin of dorsal fin slightly above middle line of trunk. Dorsal fin with about 40, anal fin with 30 supports. Each ridge-scale with three or four very prominent denticles, inclined and increasing in size backwards, flank-scales delicate, marked with more or less radiating rugæ between the pittings.

Description of Specimens.—The type specimen, originally in the collection of H. E. Strickland, exhibits the general proportions of the fish, with the characteristic stout ventral ridge-scales, but somewhat distorted in the region of the pectoral arch and lacking the greater part of the fins except the caudal. A smaller specimen in the Dorset County Museum (Pl. XIV, fig. 1) shows still better the shape of the fish as noted in the specific diagnosis above; while a larger specimen in the British Museum (Pl. XIV, fig. 2) displays the principal characters of the trunk.

In the skull (Pl. XV, fig. 1) the facial region as usual is somewhat bent downwards, while the tooth-bearing surface of the vomer is in a plane nearly parallel to that of the base of the cranium. The cranial cartilage is well ossified, one specimen in the British Museum (no. P. 1627 *a*) showing apparently the exoccipital, basioccipital, and pro-otic elements, while others seem to exhibit a postfrontal (sphenotic) and perhaps an alisphenoid or orbitosphenoid (Pl. XV, fig. 1, *ors.*). The cranial roof is completely covered with membrane bones, which are marked with a more or less radiating reticular ornament. A small narrow median element which forms the crest behind the frontals, may be described as the supraoccipital plate (Pl. XV, fig. 1, *socc.*), and bears a few close rows of small recurved denticles along its edge. It is more or less incompletely fused behind with the first ridge-scale (*r.s.*), which bears a median row of three or four recurved denticles rapidly increasing in size backwards. The lateral wings of this ridge-scale taper downwards as they pass below the limit of the supraoccipital plate. A relatively large parietal (*pa.*) bounds the supraoccipital outwardly or below, and tends also to bound the squamosal behind. It may even be interpreted as consisting of the parietal fused with a narrow supratemporal, for it is crossed by two sparse transverse rows of large openings which mark the course of slime-canals, while a long smooth digitate prominence (*x.*) from the middle of its hinder border may be a post-temporal. The shape of the bone and its reticulate ornament are well shown in an isolated specimen (Pl. XIV, fig. 4) which has unfortunately been drawn upside down. The squamosal bone, which has slipped a little beneath the parietal

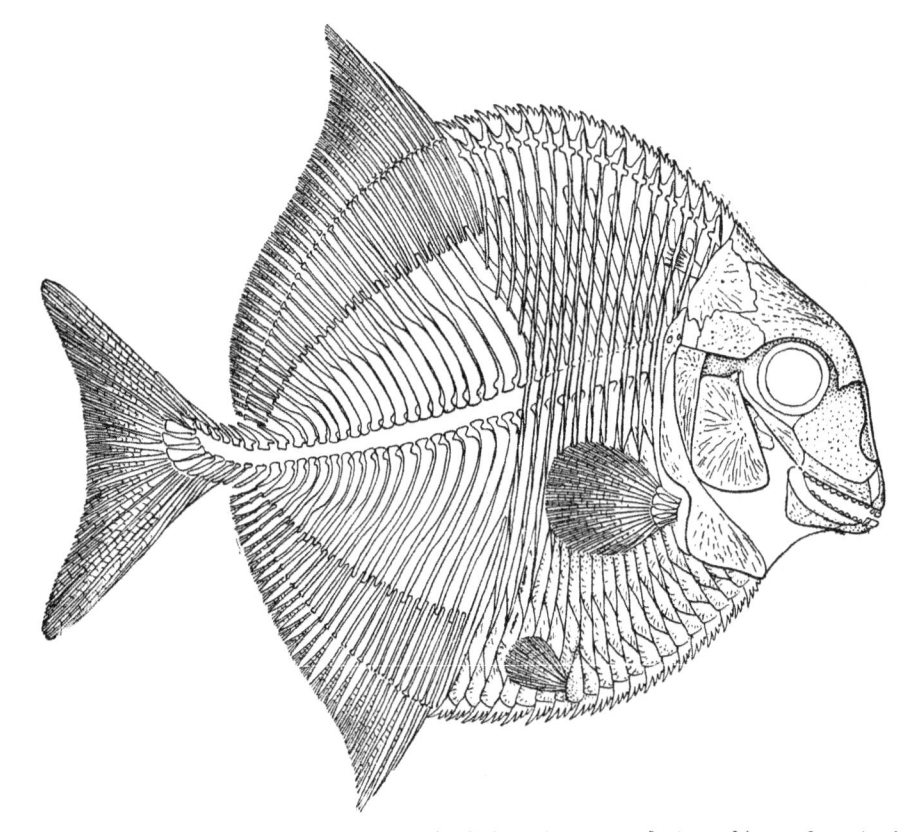

Fig. 23.—*Microdon radiatus*, Agassiz; restoration (omitting pterygo-quadrate arch), nearly nat. size.—Middle Purbeck Beds; Swanage, Dorset.

in the original of Pl. XV, fig. 1 (*sq.*), is relatively small, with its maximum width not quite equalling its maximum length. It forms the upper part of the posterior border of the orbit and completely covers the postfrontal (sphenotic). The frontals (*fr.*) are by far the largest bones of the cranial roof, widely expanded in their rather flat hinder region, tumid above the front of the orbit where they bend sharply downwards, and truncated anteriorly where they end above the middle of the anterior border of the orbit. The course of the longitudinal slime-canal is marked on their outer face by a sparse series of large openings. No superficial bones have been clearly observed in advance of the frontals covering the narrow ethmoidal region, which seems to consist of a single vertical mesethmoidal plate (*me.*) with a thickened anterior margin, which is expanded into a rounded lobe on each side in the lower half. This element is triangular in shape, ending abruptly behind within the anterior third of the orbit, and bounded below by the parasphenoid and vomer, which extend along its posterior and its anterior portion respectively. The parasphenoid (*pas.*) is relatively large, but its hinder portion beneath the cranium is obscure in all the fossils. Where crossing the orbit the bone is slender, but it bears a thin laminar vertical keel (*k.*) below, with a free wing extending forwards to the hinder border of the mesethmoid region. Where it underlies the mesethmoid in front it also expands above into a low thin vertical lamina. The single vomer (*vo.*) is widened at its oral face to support the dentition, and its upper median keel for union with the mesethmoid is low, while it extends only slightly further forwards than the latter element. No cheek-plates have been observed, but there are often traces of an ossified sclerotic ring (Pl. XIV, fig. 1).

The mandibular suspensorium is much inclined forwards to the small obliquely-cleft mouth. The hyomandibular (Pl. XIV, fig. 10, *hm.*) is a deep narrow lamina of bone, slightly curved forwards on its long axis, with a rounded expansion behind in its upper part for the support of the operculum. Its outer face in the upper half is strengthened by a curved longitudinal ridge with its concavity forwards. The quadrate and the pterygo-palatine have not been observed, but must have been comparatively delicate plates. The nature of the maxilla, if present, remains uncertain, but there are sometimes traces of a thin toothless bone flanking the vomer, which may represent a maxillary element (*e.g.*, Mus. Pract. Geol. no. 28352, and B.M. no. 44844). The premaxilla (Pl. XIV, fig. 1) is a long slender rod tapering upwards, gently arched forwards to leave a gap between itself, the vomer, and the mesethmoid, and in contact with the latter at its upper end. Its oral border bears two smooth chisel-shaped teeth, of which the inner is the larger; while its posterior or outer border seems to be notched where there may have been the narial opening. The reduced dentary bone (Pl. XIV, fig. 6) closely resembles the premaxilla in size and shape, but its two smooth chisel-shaped teeth are relatively larger. The inner of the two teeth is again the larger. This bone (Pl. XIV, fig. 3, *d.*) rests as a splint along the front margin of the splenial which forms the greater part of the mandibular ramus. The latter stout tooth-bearing bone is exposed as a wide smooth band on the outer face of the jaw below the oral border (*spl.*); and its hollowed lower portion, which would be occupied by the meckelian cartilage, is covered for the greater part of its extent by a large elongate-triangular angular plate (*ag.*).

The tritoral teeth of the vomer and splenial bones are smooth, and when unworn usually exhibit a shallow apical pit, of which the margin may be slightly crenulated (Brit. Mus. no. P. 1627 b). They are all closely arranged, and are often much worn by mastication. The vomerine dentition is only incompletely seen in the skeletons, but there can be little doubt that the large isolated example shown enlarged in Pl. XIV, fig. 5, belongs to *Microdon radiatus*. Although this specimen is much worn it still retains traces of the apical pit in some teeth, and the oral surface is gently convex from side to side. The principal median teeth are slightly more than twice as wide as long, and the alternating inner paired teeth are relatively large. The outer paired teeth are smaller and very irregular both in size and shape. The splenial dentition is well displayed in two skeletons in the British Museum (Pl. XIV, figs. 7, 8), both much worn, but retaining traces of the apical pits in the outer or lateral teeth. The principal teeth are at least as wide as the two outer series together, and the inner series of small teeth is always imperfectly developed. The teeth of the outermost series are wider than long, and much larger than those of the next series. The microscopical structure of the tritoral teeth has already been described by Owen ('Odontography,' 1840, p. 73, pl. xliii, fig. 1). Each tooth has a large pulp-cavity from which numerous very fine tubuli radiate into the ordinary dentine as far as the thick superficial layer of ganodentine.

The hyoid arch is comparatively small, and does not bear any branchiostegal rays. The epihyal (Pl. XIV, fig. 9, *eph.*) is about half as long as the ceratohyal (*ch.*), which is deepened behind and much constricted in front.

The opercular apparatus comprises solely a large preoperculum and a smaller operculum. The preoperculum, usually crushed as shown in Pl. XIV, fig. 1, is irregularly triangular in shape with a truncated apex, the maximum width of its base considerably exceeding half its depth. It is more or less firmly united to the hyomandibular, which the truncation of its apex leaves exposed (or possibly covered by a separate plate) at its upper end. Its outer face is ornamented by reticulating ridges which radiate from the middle of its anterior border; and the usual slime-canal, which traverses the anterior half of the bone, is marked in its upper portion by a slight groove, in its lower portion by a few irregularly spaced pores. The operculum (Pl. XIV, fig. 10, *op.*) is also approximately triangular in shape, but widest at its upper end and from three to four times as deep as wide. Its upper limit corresponds with that of the hyomandibular, and the ornamental reticulations on the outer face radiate from its point of suspension. When the opercular bones are removed, traces of calcified gill-supports are sometimes seen.

The vacant space for the notochord runs in a slightly sinuous curve somewhat above the middle of the trunk, and the bounding arches must have been more or less firmly united, as shown by their usually undisturbed relations in the fossils. Their triangular proximal expansions are much too small to encircle the notochord, but their characteristic anterior laminar appendages are sufficiently wide to fill at least the proximal half of the spaces between them. A few of the anterior neural spines are comparatively stout and radiating, but the other neurals, as also the hæmals in the caudal region, are nearly parallel, and the only thickening occurs again in the short hæmals at the base of the tail. None of these arches reach the dorsal or ventral border of the fish. The ribs, which are usually obscured by the overlying squamation, are also stout and bear the laminar appendages or expansions, which are especially wide in the foremost of the series (Pl. XIV, fig. 3, *r.*). Like the other arches, the ribs do not extend to the ventral border. The total number of vertebral arches is about 15 in the abdominal, 20 in the caudal region.

In the pectoral arch two membrane bones are conspicuous. The clavicle is relatively large, extending in a gentle sigmoid curve from the level of the notochordal axis to the ventral border. It is constricted at the origin of the pectoral fin, the upper part being narrow and tapering upwards, the lower part forming a large spatulate expansion, bluntly pointed below. Its smooth inner face (Pl. XIV, fig. 2, *cl.*) is concave, while its flat or slightly convex outer face (Pl. XIV, fig. 1) is marked by a feeble reticular ornament. The anterior border of its exposed sigmoidal portion seems to bear a smooth delicate laminar expansion, partly shown in Pl. XIV, fig. 2, but better seen in Brit. Mus. no. 44844. The supraclavicle (Pl. XIV, fig. 10, *scl.*) is a deep and narrow bone crossed as usual at its upper end by the slime-canal of the lateral line. It is widest and thickest superiorly where a triangular area with a postero-inferior extension is exposed and marked with a reticulate ornament. Its lower covered portion tapers to a slender point. The cartilages of the pectoral arch are unknown, but the pectoral fin is lobate and supported by a few long and stout hour-glass-shaped basals (Pl. XIV, fig. 1). Its rays are also broad and closely subdivided. The pelvic fins, of which one is seen slightly displaced in Pl. XIV, fig. 2, *plv.*, are relatively small, inserted about midway between the pectorals and the anal, and comprise five or six very broad rays which are articulated and subdivided. The rays of the median fins are also rather broad, with numerous articulations and some subdivision. There are slightly more than 40 rays in the dorsal fin, slightly more than 30 rays in the anal fin, and in both cases the few foremost supports are the longest and most crowded to bear the long anterior rays. Unfortunately these fins are not well seen in any known specimen. The characteristic forked caudal fin, with its constricted pedicle, is well shown in Pl. XV, fig. 2. A few of its foremost basal rays both above and below are in the form of long and slender uniserial fulcra, but most of its rays, about 18 in number, are articulated with step-shaped joints and also subdivided distally. All the fin-rays are smooth.

The peculiar rod-shaped bone apparently bounding the abdominal cavity behind (noticed in Catal. Foss. Fishes Brit. Mus., pt. iii, p. 196) is sometimes recognisable above the hindmost ventral ridge-scale (Pl. XIV, fig. 1, *x.*).

As shown by the ridge-scales, the total number of transverse series of scales covering the abdominal region is about 16, but they are represented for the greater part of their extent solely by the thickened anterior rim. The ridge-scales, which are arranged in a regular close series, are much thickened, and each bears a row of three large, hollow, smooth denticles increasing in size backwards, sometimes with an additional one or two minute denticles in front. The massive smooth ridge-scales of the ventral border exhibit wide facettes for overlap by the lowest flank-scales (Pl. XIV, figs. 1 *b*, 2); while each of those of the dorsal border is produced downwards into a triangular lateral wing, which is more or less ornamented by reticulations like the other scales. The flank-scales are complete only in the lower part of the abdominal region, diminishing in number and vertical extent backwards. As shown by a specimen from Ridgeway, in the Dorset County Museum, there are five of these scales in at least one anterior transverse series, four scales in several following series, then three scales, and finally only two complete scales in at least one series adjacent to the origin of the anal fin. All these scales (Pl. XV, fig. 3) are much deeper than wide, increasing in depth upwards, and the uppermost scale ends in an attenuated apex. The line of union between each scale and the next above is very oblique, and the posterior border is gently convex. The outer surface is ornamented with reticulations which tend to radiate from the middle of the smooth anterior margin. This margin has a wide thickening on the inner face, which forms the usual deep peg-and-socket articulation (Pl. XV, figs. 4, 5). At the position of the pelvic fins, which are inserted in a triangular hollow on the flank just above the ventral border, the lowest scales of two transverse series diverge to allow the intercalation of a deep triangular scale, and one of the ridge-scales is extended to bear a second facette for the overlap of this additional scale (as shown in Pl. XIV, fig. 2, *dr.*). The lateral line passes through a perforation in the degenerate riblet of at least the two anterior series of scales; and all the riblets immediately below the dorsal ridge-scales are slightly widened and pierced along the course of the upper lateral line. A single smooth, nearly rhombic scale is also sometimes seen at the base of the caudal fin, perhaps the last remnant of the squamation of an upper caudal lobe.

The principal characters of the species, as now described, are shown in the restoration of the fish, Text-fig. 23, p. 60.

Horizon and Localities.—Middle Purbeck Beds: Swanage and neighbourhood of Weymouth, Dorset.

Genus COELODUS, Heckel.

Cœlodus, J. J. Heckel, Sitzungsb. k. Akad. Wiss. Wien, math.-naturw. Cl., vol. xii, 1854, p. 455.
Glossodus, O. G. Costa (*non* McCoy, 1848), Atti Accad. Pontan., vol. vii, 1853, p. 26.
Cosmodus, H. E. Sauvage, Bull. Soc. Sci. Nat. Yonne, vol. xxxiii, pt. ii, 1879, p. 48.

CŒLODUS.

Generic Characters.—Trunk deeply fusiform, with short slender caudal pedicle. Cranial roof-bones pierced by a pair of supratemporal vacuities, surrounded by the supraoccipital, parietal, and frontal elements. Head and opercular bones externally rugose and punctate; two chisel-shaped teeth in each premaxilla and dentary; tritoral teeth often exhibiting an apical indent usually with crenulated border; oral surface of vomer convex from side to side, with teeth in five longitudinal series; splenial dentition comprising three series of teeth with long axes more or less directly transverse, sometimes supplemented within by a small row. Neural and hæmal arches of axial skeleton of trunk not expanding sufficiently to encircle the

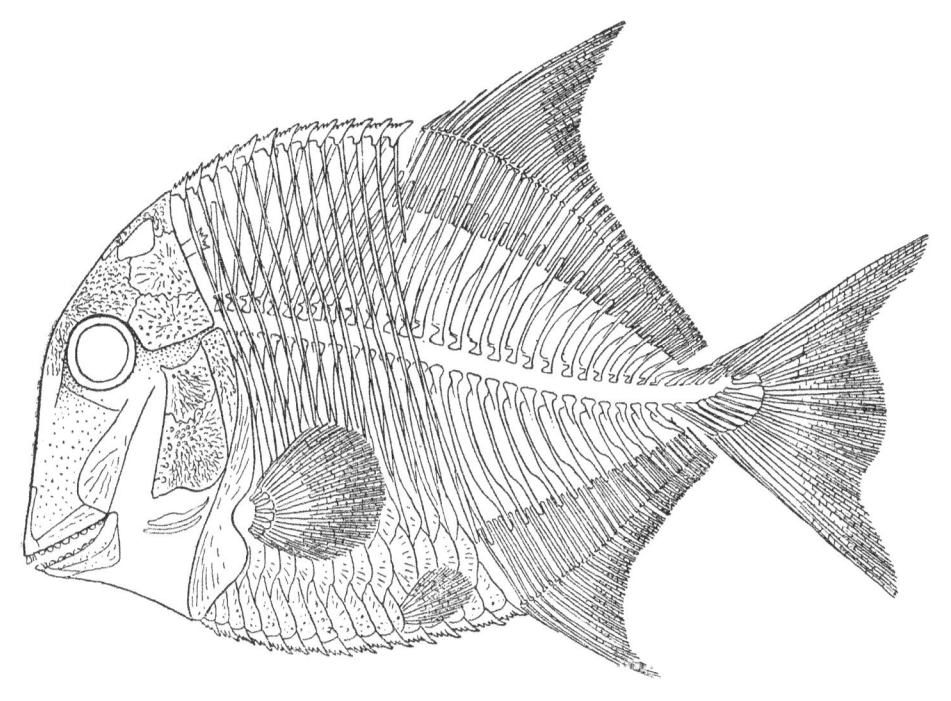

Fig. 24.—*Cœlodus costæ*, Heckel; restoration (omitting pterygo-quadrate arch), nearly nat. size.—Lower Cretaceous; Torre d'Orlando, Castellamare, Naples.

notochord. Fin-rays robust, closely articulated, and much divided distally. Pelvic fins present; dorsal and anal fins high and acuminate in front, low and fringe-like behind, the former occupying at least the hinder half of the back, and the latter somewhat shorter, arising more posteriorly; caudal fin forked, with a convexity in the middle. Scales ornamented with reticulating rugæ and punctations, occupying only the anterior half of the trunk in advance of the median fins, and complete only in the ventral region.

Type Species.—*Cœlodus saturnus* (Heckel, *loc. cit.*, 1854, p. 455, and Denkschr. k. Akad. Wiss. Wien, math.-naturw. Cl., vol. xi, 1856, p. 207, pls. iii, iv), known by a nearly complete fish from the Lower Cretaceous of Goriansk, Istria.

Remarks.—This is typically a Cretaceous genus and is known by several complete skeletons; but only jaws have hitherto been found in the Wealden and

Purbeck formations. The dentition is so variable that such fossils can only be provisionally named.

A restoration of the skeleton of a small species, *Cœlodus costæ*, based partly on specimens in the British Museum (nos. P. 1671, P. 1671 a, P. 4394), partly on figures by Bassani and D'Erasmo,[1] is given in Text-fig. 24.

1. **Cœlodus mantelli** (Agassiz). Plate XV, figs. 6—11.

1827. "Palates of an unknown fish," G. A. Mantell, Foss. Tilgate Forest, p. 58, pl. xvii, figs. 26, 27.
1833. *Pycnodus microdon*, L. Agassiz, Poiss. Foss., vol. ii, pt. i, p. 17.
1839–44. *Pycnodus mantellii*, L. Agassiz, tom. cit., pt. ii, p. 196, pl. lxxii a, figs. 6—14.
1839–44. *Gyrodus mantellii*, L. Agassiz, tom. cit., pt. ii, p. 234, pl. lxix a, fig. 18.
1853. *Glossodus mantellii*, O. G. Costa, Atti Accad. Pontan., vol. vii, p. 28.
1856. *Cœlodus mantelli*, J. J. Heckel, Denkschr. k. Akad. Wiss., math.-naturw. Cl., vol. xi, p. 203.
1895. *Cœlodus mantelli*, A. S. Woodward, Catal. Foss. Fishes B.M., pt. iii, p. 252.

Type.—Jaws; British Museum.

Specific Characters.—A species of small or moderate size known only by the jaws and dentition which rarely exceed 2 cm. in length. All teeth smooth, with a shallow apical pit, which in unworn specimens is faintly crenulated round the margin. Teeth of median series on vomer somewhat more than twice as broad as long, with a concave posterior margin, sometimes mesially constricted; teeth of two lateral series nearly equal in size, often slightly elongated antero-posteriorly, their width together not equalling that of the median series. Teeth of principal series on the splenial bone about twice as broad as long, scarcely tapering inwards, slightly raised at their outer end; their width scarcely equalling that of the two outer series, of which the innermost is nearly twice as wide as the outermost; the inner of these teeth more or less raised at the inner end; a series within the principal series rarely represented even by scattered small teeth. Initial anterior end of splenial narrow, with the teeth in three regular series.

Description of Specimens.—The original portions of dentition in the Mantell Collection, described by Agassiz, are rightly identified as belonging both to the vomerine and splenial bones, and exhibit most of the principal characters of the species. One specimen showing the narrow splenial dentition of a young individual, with three regular series of teeth, is also described under the name of *Gyrodus mantellii*. All these fossils were obtained from the Wealden (presumably Tunbridge Wells Sands) of Tilgate Forest, Sussex.

The largest known specimen of the vomerine dentition, already figured by Agassiz, tom. cit., pl. lxxii a, fig. 12, is shown enlarged in Pl. XV, figs. 6, 6 a–c. The median teeth are well spaced and all are worn by mastication except the

[1] F. Bassani and G. D'Erasmo, "La Ittiofauna del Calcare Cretacico di Capo d'Orlando presso Castellammara (Napoli)," Mem. Soc. Ital. Sci. [3], vol. xvii (1912), pl. v, figs. 4, 5.

hindmost, which displays its median constriction, raised ends, and the transversely elongated apical pit with a crenulated anterior rim. To the left the lateral teeth are covered by hard matrix, but on the right side (fig. 6 a) both series are well seen, with traces of small intercalated teeth in front; all probably had an apical pit, but they are so much worn that only the hindmost tooth of the outer series retains the crenulated rim. The teeth in smaller examples of the vomerine dentition have a larger and deeper apical pit with crenulated margin; but the specimens are usually so much worn by mastication and accidentally flaked in the fossils, that this feature is often obscured. In many cases the wearing of the teeth reaches the pulp-cavity, and in one specimen figured by Agassiz (*tom. cit.*, pl. lxxii a, fig. 13) only the broken bases of insertion are seen on the bone.

In specimens of the adult splenial dentition, as in one figured by Agassiz (*tom. cit.*, pl. lxxii a, fig. 14) and in the original of Pl. XV, fig. 7, the anterior tapering end is usually broken away; but some of the immature stages represented by this end are shown in Pl. XV, figs. 8—11. The smallest of these specimens (fig. 9) was wrongly referred by Agassiz (*loc. cit.*) to the upper jaw of *Gyrodus*, under the name of *G. mantellii*. It is a left splenial bone, of which the inner margin is recognisable in the matrix, and it exhibits the three characteristic regular rows of teeth, each with a large apical pit which has been more or less reduced by wear. A larger left splenial (fig. 10) bears teeth which are not only much worn but also a little flaked. The third small splenial dentition (fig. 8) has already been figured by Agassiz (*tom. cit.*, pl. lxxii a, fig. 11), and is interesting as still retaining the crenulations round the margin of the apical pit in the hinder teeth. The fourth specimen (fig. 11), which is detached from the matrix, shows well (fig. 11 a) the depth of the groove in the dental armature formed by the lateral teeth, and the extreme wear in the front part of this groove. Several small examples of the dentition from the Wealden (presumably Weald Clay) of Sevenoaks, Kent, in the Museum of Practical Geology, are interesting both for the fine unworn condition of many of the teeth and for the irregular subdivision of some others.

Horizons and Localities.—Wealden: Tilgate Forest, Tunbridge Wells, and Hastings, Sussex; Atherfield, Isle of Wight; Sevenoaks, Kent. Lower Wealden or Upper Purbeck: Netherfield, Sussex.

2. **Cœlodus multidens,** sp. nov. Plate XV, figs. 12, 13.

Type.—Splenial bone; Museum of Practical Geology, London.

Specific Characters.—Initial anterior end of splenial bone wider than in *C. mantelli*, covered with an irregular group of smooth rounded teeth, each with an apical pit; mature splenial teeth almost as in *C. mantelli*, but apical pit apparently shallower.

Description of Specimens.—This species is known only by the splenial bone, of which the anterior part bears teeth like those of *Athrodon*. The normal dentition does not begin until the bone has attained a width of at least a centimetre. The type specimen (Pl. XV, fig. 12) exhibits the short and broad bone with a wide inner border free from teeth. The irregular teeth occupy more than half of its length, but the hinder teeth are almost exactly those of *Cœlodus mantelli*. All are much worn except the two teeth most posteriorly, which are not crimped round the shallow apical pit. In part of a smaller specimen (Pl. XV, fig. 13) the teeth are extremely worn by mastication, and the principal teeth are inclined at a considerable angle to those of the outer series.

Horizon and Localities.—Wealden: Sevenoaks, Kent; Battle, Sussex; Brook and Atherfield, Isle of Wight.

3. **Cœlodus hirudo** (Agassiz). Plate XV, figs. 14—18.

1839. *Acrodus hirudo*, L. Agassiz, Poiss. Foss., vol. iii, p. 148, pl. xxii, fig. 27.
1887. ,, ,, A. S. Woodward, Geol. Mag. [3], vol. iv, p. 102.
1889. ,, ,, A. S. Woodward, Catal. Foss. Fishes B. M., pt. i, p. 296, pl. xiii, fig. 9.

Type.—Much worn tooth; British Museum.

Specific Characters.—A species known only by isolated dental crowns which sometimes attain a length of 2·5 cm. in longest diameter, but are usually smaller. Principal teeth slightly more than twice as broad as long; coronal contour gently rounded, somewhat raised near each lateral end, and the surface marked by very fine wrinkles diverging or radiating from an extended apical pit which is reduced to an inconspicuous groove.

Description of Specimens.—The type specimen, which was misunderstood by Agassiz, is shown from the upper, anterior, and lateral aspects in Pl. XV, figs. 14, 14 *a, b*, and may be regarded as a principal tooth of the left splenial bone. It is imperfect at the outer and posterior borders, and its upper surface (fig. 14) has been removed by mastication, which has produced a deep hollow in the outer half. The posterior broken edge shows that the main part of the tooth consists only of a comparatively thin crown, without any root. The anterior face (fig. 14 *a*), which is described and figured by Agassiz as the top of the crown, retains the characteristic ornament of very fine vertical wrinkles. Concentric lines of growth are conspicuous round the base of the crown, which is shown to have been fixed to the bone by a peripheral root in the usual Pycnodont manner.

A smaller tooth evidently of the same type was described and figured *loc. cit.* 1889, displaying the shape and ornament of the unworn crown; and another good specimen of intermediate size is shown in Pl. XV, figs. 15, 15 *a, b*. Here the oral surface (fig. 15) is only slightly worn, so that the original contour of the tooth is obvious. The anterior as well as the posterior face (fig. 15 a) exhibits traces of concentric lines of growth across the characteristic wrinkled ornament, and the peripheral nature of the root is clear (figs. 15 a, b). A smaller specimen (fig. 16) shows well the remnant of the apical pit as a slight fissure from which the coronal wrinkles diverge.

The lateral teeth have not been identified with certainty, but the low-crowned ovoid specimen represented in Pl. XV, fig. 17, may be one of them. The peripheral ornament of the crown resembles that of the principal teeth of this species, but that of the middle of the crown is a coarse irregular tuberculation. Among other teeth that are evidently Pycnodont, the small elongated specimen with a high crown shown in Pl. XV, fig. 18, may also belong to one of the lateral series of *C. hirudo*.

A microscopical section made from one of the principal teeth exhibits the typical Pycnodont structure, densely arranged minute tubuli traversing the dentine from the pulp-cavity direct to the superficial ganodentine.

Horizon and Localities.—Wealden: Tilgate Forest; neighbourhood of Hastings (Wadhurst Clay and Ashdown Sands); Telham, near Battle.

4. **Cœlodus lævidens**, sp. nov. Plate XV, figs. 19, 20.

Type.—Splenial bone; British Museum.

Specific Characters.—A small species with splenial dentition usually not more than 2·5 cm. in length. Teeth of principal series on the splenial bone about twice as broad as long, scarcely tapering inwards and not much (if at all) raised at the outer end; surface smooth and apical pit absent or very shallow; the width equalling or somewhat exceeding that of the two flanking series. Teeth of inner flanking series also about twice as broad as long, only slightly raised at the inner end, smooth but with a deeper apical pit. Teeth of outer flanking series smaller, less transversely elongated, smooth but with shallow apical pit.

Description of Specimens.—The type specimen (Pl. XV, fig. 19) is the hinder part of a right splenial, of which the hindmost principal tooth is scarcely worn but still shows no more than the feeblest trace of an apical pit. The two hindmost flanking teeth, though pitted, are not crimped. The younger specimen of the left splenial shown in Pl. XV, fig. 20, is altogether similar, and bears the three rows of teeth in regular series to the anterior apex. Though in most respects resembling the corresponding dentition of *Cœlodus mantelli*, the absence of a well-defined apical pit in the principal teeth and the shape of the inner flanking teeth are characters suggesting that this form of jaw belongs to a distinct species.

A vomerine dentition closely similar to that of *C. mantelli* is known from the same horizon as the splenial just described, and probably belongs to this species.

Horizon and Locality.—Middle Purbeck Beds: Swanage, Dorset.

5. **Cœlodus arcuatus**, sp. nov. Plate XIII, fig. 5.

Type.—Vomerine dentition; Museum of Practical Geology, London.

Specific Characters.—A species known only by the vomerine dentition, which measures about 1·5 cm. in maximum width. Teeth of median series about three times as broad as long, much constricted mesially, and much arched backwards; apical pit well marked, crenulated round the margin, and the hinder face of the tooth vertically plicated in its concavity. Teeth of two lateral series nearly equal in size, and their width together not equalling that of the median series; apical pit well marked, large, and crenulated round the margin.

Description of Specimen.—The only known example of the vomerine dentition (Pl. XIII, fig. 5) is imperfect anteriorly but otherwise beautifully preserved. It exhibits the usual transverse convexity, but the teeth of the paired series are on a less steeply sloping surface than those in *C. mantelli*. The plication of the posterior concavity of the median teeth is especially characteristic of the species.

Horizon and Locality.—Middle Purbeck Beds: Swanage, Dorset.

Family MACROSEMIIDÆ.

Genus OPHIOPSIS, Agassiz.

Ophiopsis, L. Agassiz, Neues Jahrb. f. Min., etc., 1834, p. 385; and Poiss. Foss., vol. ii, pt. i, 1844, p. 289.

Generic Characters.—Trunk much elongated, and the dorsal margin only slightly arcuate; head large or of moderate size. Marginal teeth acutely pointed. Notochord surrounded by delicate ring-vertebræ; ribs ossified. Bifurcation of dorsal fin-rays variable; fulcra present, comparatively stout at the base of the dorsal and caudal fins. Paired fins relatively large; dorsal fin ordinarily extending about half the length of the back, high in front, low behind; anal fin small and well forwards; caudal fin forked. Scales covering the whole of the trunk, in regular series, united by peg-and-socket articulation, and often pectinated at the hinder border; the scales of the middle of the flank scarcely deeper than broad, few of the ventral scales much broader than deep; no enlarged ridge-scales.

Type Species.—*Ophiopsis procera* (L. Agassiz, Poiss. Foss., vol. ii, pt. i, 1844, p. 289, pl. xlviii, fig. 1) from the Lower Kimmeridgian (Lithographic Stone) of Bavaria.

Remarks.—The restoration of the type species of *Ophiopsis* given in Text-fig. 25 is based chiefly on a specimen in the British Museum (no. P. 6939), in which all the fins except the anal are nearly complete. In this fossil the ring-vertebræ are sufficiently stout not to have collapsed by crushing, and thus form a ridge beneath the squamation. The pectoral fin is strengthened at the base of its foremost ray by two or three large fulcra which rapidly increase in length backwards. The pelvic fin is relatively large and fringed with conspicuous fulcra. The dorsal fin, which rises to a sharp eminence in front, has five or six basal fulcra increasing in length; and its fringing fulcra do not extend to the distal end of the foremost ray. The position of the very small anal fin is indicated, and this fin is restored from other specimens. The fork of the caudal fin is well shown. The large postclavicular scales are nearly smooth, but apparently serrated at their hinder border. Some of the anterior flank-scales are not only serrated, but also

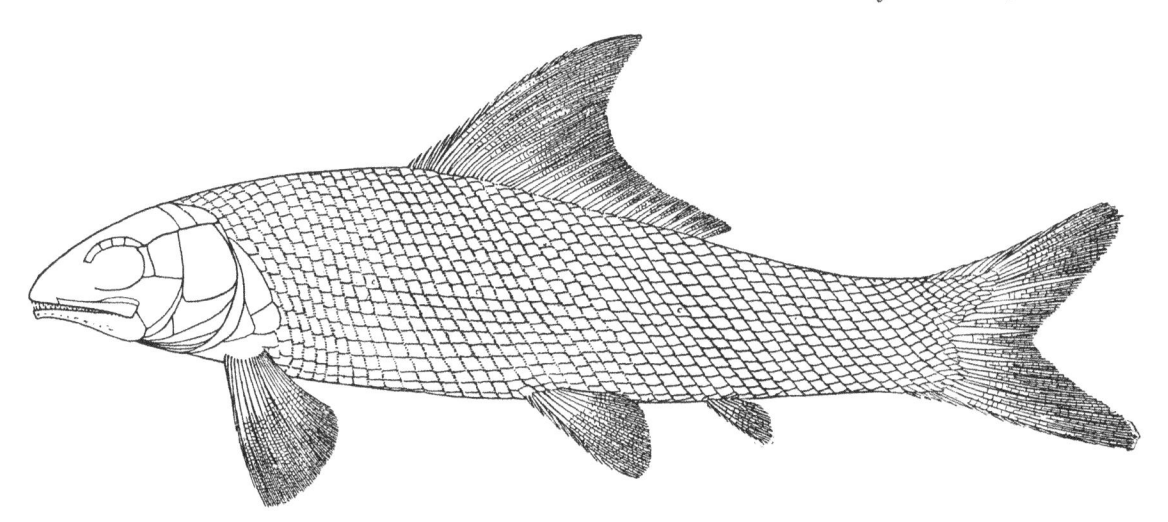

FIG. 25.—*Ophiopsis procera*, Agassiz; restoration, about two-thirds nat. size.—Lower Kimmeridgian (Lithographic Stone); Bavaria. Based chiefly on a specimen in the British Museum (no. P. 6939).

marked with pectinations extending from the serrations. The lateral line is inconspicuous, only indicated by an occasional short vertical slit or a postero-inferior notch on the scales which it traverses. Along the base of the dorsal fin, small irregular triangular scales are intercalated at the upper ends of the transverse series. The only enlarged ridge-scale is a flat oval scale at the beginning of the fulcral series on the dorsal border of the caudal pedicle.

1. **Ophiopsis penicillata,** Agassiz. Plate XVI, figs. 1, 2.

1844. *Ophiopsis penicillata*, L. Agassiz, Poiss. Foss., vol. ii, pt. i, p. 290, pl. xxxvi, figs. 2—4.
1895. *Ophiopsis penicillata*, A. S. Woodward, Catal. Foss. Fishes B.M., pt. iii, p. 169.

Type.—Nearly complete fish; British Museum.
Specific Characters.—A robust species about 18 cm. in length. Length of head

with opercular apparatus somewhat exceeding maximum depth of trunk, which is twice as great as depth of caudal pedicle and contained about five times in total length of fish. External head-bones coarsely tuberculated or rugose; marginal teeth long and slender. Dorsal fin arising at the end of the anterior third of the back, half as long as the trunk, comprising not less than 25 rays, mostly bifurcated, of which the longest do not equal the depth of the trunk at their point of insertion; pelvic fins arising slightly in advance of the middle point between the pectorals and the caudal. Scales smooth, with finely serrated hinder border except towards the hinder end of the caudal region where they are entirely smooth.

Description of Specimens.—The type specimen (Pl. XVI, fig. 1) is contained in Purbeck stone of uncertain origin, and exhibits most of the specific characters of the fish. Though not much distorted, it is considerably fractured by crushing; and it is represented in a rather diagrammatic manner in the original drawing published by Agassiz. The head is probably not lengthened by distortion, but the cranial roof is pushed over to the left side, so that the right squamosal and some other bones are seen from below. One fragment of bone displays the coarse ornament of tubercles of ganoine, which are partially fused together. Slender styliform teeth, with a sharply pointed and somewhat incurved apex (fig. 1 *a*), occur in both jaws. The left operculum, exposed from within, is about two-thirds as wide as deep. Below it are the comparatively small suboperculum and interoperculum, and some remains of branchiostegal rays. Above it may be recognised the left side of the pair of supratemporal plates, and a much larger post-temporal which is in contact with an elongated supraclavicle, passing below to the much-arched smooth clavicle. The axial skeleton of the trunk must have been imperfectly ossified, but there are traces of vertebral centra and stout neural arches just behind the occiput, and more distinct remains (though not so clear as figured by Agassiz) also occur in the hinder half of the tail. The caudal vertebral rings seem to have been longer than deep, with stout neural and hæmal arches; and the upturned end of the vertebral axis is distinct. The two pectoral fins are crushed together, each probably consisting of about twelve rays, which are closely articulated and bifurcating in more than their distal half. They do not exhibit fulcra. The pelvic fins, also crushed together, are much smaller, and seem to bear a few slender fulcra. The extended dorsal fin shows about five basal fulcra in front, and its well-spaced stout rays are all articulated and bifurcated distally. The complete length of its anterior rays seems to be exhibited. The anal fin is represented by a mere fragment, with some of its supports. The large and strong caudal fin, with slender fulcra along its upper border, must have been forked. The greater part of the squamation of the left side is undisturbed and exposed from its inner face, while traces of the scales of the right side, in outer view, are seen near the ventral margin. There are enlarged post-clavicular scales, at least just above the insertion of the pectoral fin. None of the flank-scales seem to have been deeper than broad, most of the scales being broader than deep, and those near the ventral border especially so. On all the scales the inner vertical rib is feebly marked and wide, and in the principal scales (fig. 1 b) the peg-and-socket articulation behind this rib is wide and shallow. The caudal scales (fig. 1 c) are not united by peg-and-socket. In the abdominal region the lateral line traverses about the eighth row from the dorsal, the twelfth or thirteenth row from the ventral border. The exhibited outer face of the ventral scales is smooth and slightly concave, and the hinder margin seems to have been feebly serrated. There are no enlarged ridge-scales.

A more imperfect and slightly larger specimen obtained by Mr. R. F. Damon from the Purbeck Beds of the Isle of Portland (Brit. Mus. no. P. 8375), evidently belongs to the same species and shows a few additional features. The premaxilla, as usual in *Ophiopsis*, is relatively large and much expanded. The vertebræ in the abdominal region are complete cylinders about as long as deep, and the short ribs are remarkably slender. Each of the stout rays of the dorsal fin bears a small and slender pointed prominence directed backwards at its lower or articular end (Pl. XVI, fig. 2). The scales are smooth, and some of them exhibit very fine and delicate serrations on the hinder border.

Another imperfect fish of the same species from the West Quarry, Ridgeway, near Weymouth, now in the Dorset County Museum, shows the small upper circumorbital plates ornamented with large flattened tubercles of ganoine. A displaced fragment of bone bearing minute teeth seems to be part of the splenial.

Horizon and Locality.—Lower Purbeck Beds: near Weymouth, Dorset.

2. **Ophiopsis breviceps**, Egerton. Plate XVI, figs. 3—12.

1852. *Ophiopsis breviceps*, P. M. G. Egerton, Figs. and Descript. Brit. Organic Remains (Mem. Geol. Surv.), dec. vi, no. 6, pl. vi.
1895. " " A. S. Woodward, Catal. Foss. Fishes B. M., pt. iii, p. 170.

Type.—Nearly complete fish; Museum of Practical Geology, London.

Specific Characters.—A robust species about 12 cm. in length. Length of head with opercular apparatus about equal to maximum depth of trunk, which is twice as great as depth of caudal pedicle and contained about four and a half times in total length of fish. External head-bones coarsely tuberculated or rugose; marginal teeth long and slender. Dorsal fin occupying greater part of hinder two-thirds of back, with about 35 rays, of which the longest do not equal the depth of the trunk at their point of insertion; pelvic fins arising almost at the middle point between the pectorals and the caudal. Scales smooth and somewhat concave externally, with a strongly but finely serrated hinder border.

Description of Specimens.—The type specimen (Pl. XVI, fig. 3), though deepened a little by crushing, shows the general proportions of the fish and the origin of all the fins. The jaws of the left side, seen from within, are in their natural position; but the broken roof of the skull is displaced upwards and exposed from beneath. The calcified vertebral rings, though crushed and broken, are distinct; and the scales, somewhat scattered, are well displayed both in outer and in inner view. Though imperfect, the remains of the anal fin prove that it must have been comparatively small.

More or less fragmentary examples of this species are common in the Lower Purbeck Beds near Tisbury, Wiltshire, and the scattered remains are interesting as showing well several osteological characters of the fish. All the external bones appear to have been ornamented with large flat tubercles of ganoine, which are often variously fused into a vermiculating pattern (Pl. XVI, fig. 7), sometimes into a continuous film. As seen from below in one specimen (Pl. XVI, fig. 4, *pa.*) the parietal bones are longer than wide; and as seen also from above in another specimen (B. M. no. P. 9107 *b*), they are united in a very wavy median suture. The frontal bones (*fr.*) are slightly more than twice as long as the parietals, excavated laterally by the large orbit, and ending in front rather bluntly. The external ornament seems to be confined to their posterior half (B. M. no. P. 9436). A long and narrow squamosal (*sq.*), bearing an extended hyomandibular facette, bounds the parietal on each side. The parasphenoid is relatively large and expanded (B. M. no. P. 9107 *b*). The cheek is covered both by postorbital and by circumorbital plates. The upper postorbital is relatively large, deeper than wide, widest below. The large lower postorbital exhibits a few short branches radiating backwards from the curved slime-canal (Mus. Pract. Geol. no. 28441). The circumorbital ring is narrow, including three or four plates above the orbit (*co.*) and one behind. These plates are marked with the usual coarse ornament (B. M. no. P. 9436). The hyomandibular is a deep and narrow lamina of bone, with a large process for the support of the operculum. The maxilla seems to have been long and slender, but its precise shape is unknown. The premaxilla (Pl. XVI, fig. 5) forms a large irregular expansion, pierced near its upper border by a small foramen, and produced upwards into a short and slender ascending process. The oral margin bears a row of styliform teeth smaller than those of the dentary. The lower jaw (Pl. XVI, fig. 6) is much elevated in the short coronoid region, but the tooth-bearing part of the dentary is long and slender. Its outer face is marked by a row of large openings of the slime-canal, and its slender styliform teeth are arranged in a close series. There seems to have been a splenial with minute teeth (B. M. no. P. 3608). The occiput is overlapped behind by a single pair of supra-temporals as in *Amia* (Pl. XVI, fig. 4, *st.*). The operculum is not much deeper than wide, and its inner face bears a large facette for its suspension; the coarse ornament on its outer face (Pl. XVI, fig. 7) tends to radiate from this point. The suboperculum is nearly three times as wide as deep, with an ascending process in front. The uppermost branchiostegal ray is relatively large, and the others are also stout. There seems to have been a large gular plate, but its presence is not quite certain.

In the axial skeleton of the trunk complete vertebral rings, about as long as deep, occur throughout; in the abdominal region the ribs are short and slender (B. M. no. P. 9107 c).

Behind the supratemporal bones, a pair of relatively large post-temporals is seen (Pl. XVI, fig. 4, *ptt.*). The supraclavicle is also large (Pl. XVI, fig. 8), about three times as deep as wide, with its exposed surface ornamented by ganoine partly disposed in irregular oblique ridges. The clavicle is wide and much arched, and similarly ornamented on its small exposed portion (B. M. no. P. 9436). The large upper postclavicular scale is shown in the type specimen (Pl. XVI, fig. 3). Fulcra have been seen on the pelvic fins (B. M. no. P. 9107 c), and there seem to be traces of them on the pectorals in the type specimen.

All the scales are smooth, and those of the abdominal region, besides many of the caudal region, are conspicuously though finely serrated. The principal scales of the abdominal flank are about as broad as deep, with gently curved upper and lower margins, and often marked with a few faint zig-zag lines parallel with the posterior serrations (Pl. XVI, fig. 9). They are united by a large peg-and-socket articulation (fig. 10), which becomes feeble or absent on the dorsal and ventral scales and in the hinder half of the caudal region. The dorsal scales are much broader than deep, with a tendency to the rounding of the postero-superior angle (fig. 12). The ventral scales are also broader than deep (fig. 11), those in the anterior half of the abdominal region excessively so. Ovate ridge-scales occur on the caudal pedicle, but they are scarcely enlarged.

Horizon and Locality.—Lower Purbeck Beds: Vale of Wardour, Wiltshire.

3. Ophiopsis dorsalis, Agassiz. Plate XVI, fig. 13.

1844. *Ophiopsis dorsalis*, L. Agassiz, Poiss. Foss., vol. ii, pt. i, p. 291, pl. xxxvi, fig. 5.
1895.   „   „   A. S. Woodward, Catal. Foss. Fishes B. M., pt. iii, p. 171.

Type.—Nearly complete fish; British Museum.

Specific Characters.—A much elongated species about 16 cm. in length. Head with opercular apparatus occupying one-fifth of the total length of the fish; maximum depth of trunk twice as great as depth of caudal pedicle, and contained somewhat more than six times in the total length. External head-bones coarsely tuberculated or rugose. Dorsal fin occupying greater part of back, with about 35 rays, of which the longest do not equal the depth of the trunk at their point of insertion; pelvic fins in advance of the middle point between the pectorals and caudal. Principal scales smooth and somewhat concave externally, with a finely serrated hinder border; some scales irregularly punctated or rugose.

Description of Specimens.—The type specimen, which is very imperfectly shown in the original figure published by Agassiz, is re-drawn of the natural size in Pl. XVI, fig. 13. The outline of the head is indicated, and there are traces of the coarse ornamentation, but the details of its osteology are obscure. The shape of the trunk is also clearly shown, and the fins are not very imperfect. There are no indications of fulcra on the pectoral fins, but they are distinct on the pelvic fins; both basal and fringing fulcra are also seen on the dorsal fin, and they are conspicuous on the upper lobe of the caudal fin. The anal fin is comparatively small and deep, with eight or nine rays. The comparatively stout caudal fin-rays are well enamelled. The scales are exposed chiefly from the inner face, and exhibit the variations in shape already described in *O. breviceps* (p. 75). Some of the scales in the abdominal region exhibit their fine serration, while a few on the caudal pedicle are covered with faintly rugose enamel. The lateral line in the abdominal region traverses the ninth or tenth row above the ventral border.

The type specimen is described by Agassiz as from the Inferior Oolite of Northampton, but the matrix appears to be Purbeck Stone, and a second example of the same species in the British Museum was certainly obtained from the Purbeck Beds of Swanage. In this specimen a few of the scales below the middle of the dorsal fin, and just in front of the pelvic fins, exhibit the faint rugosity of the enamel already noted on the caudal pedicle of the type. There is another example from Swanage in the Museum of Practical Geology, London.

Horizon and Locality.—Middle Purbeck Beds: Swanage, Dorset.

Genus **HISTIONOTUS**, Egerton.

Histionotus, P. M. G. Egerton, Ann. Mag. Nat. Hist. [2], vol. xiii, 1854, p. 434.

Generic Characters.—Head large, snout acute; dorsal margin of trunk rising above the head to an angulation from which the body gradually tapers backwards. Styliform marginal teeth very slender and closely arranged. Slime-canals on head and preoperculum comparatively large. Notochord surrounded by ring-vertebræ; ribs ossified. Fins consisting of distally bifurcating rays, all with Λ-shaped fulcra; pectoral fins much larger than the pelvic pair; dorsal fin arising at the angulation of the back, extending to the caudal pedicle, high in front, low behind; anal fin small; caudal fin forked. Scales covering the whole of the trunk, in regular series, united by peg-and-socket articulation, and more or less pectinated at the hinder border; scales of middle of flank and some of dorsal region deeper than broad, the flank-scales with more or less convex hinder border; ventral scales at least as broad as deep; postclavicular scales very large; caudal ridge-scales not much enlarged.

Type Species.—*Histionotus angularis*, from the English Purbeck Beds.

Remarks—*Histionotus* is noteworthy for the great development of the slime-canals on the head and preoperculum. It is known only by the type species and by two or three others from the Lithographic Stone (Lower Kimmeridgian) of Bavaria and France.

1. Histionotus angularis, Egerton. Plate XVII, figs. 1—5.

1854-55. *Histionotus angularis*, P. M. G. Egerton, Ann. Mag. Nat. Hist. [2], vol. xiii, p. 434, and Figs. and Descript. Brit. Organic Remains (Mem. Geol. Surv.), dec. viii, no. 5, pl. v.
1889. *Histionotus angularis*, J. C. Mansel-Pleydell, Geol. Mag. [3], vol. vi, p. 241, pl. vii.
1895. *Histionotus angularis*, A. S. Woodward, Catal. Foss. Fishes B.M., pt. iii, p. 174.

Type.—Fish, wanting tail; British Museum.

Specific Characters.—Attaining a length of about 20 cm. Length of head with opercular apparatus slightly exceeding its maximum depth, and occupying about one-quarter of the total length of the fish; length of trunk equalling twice its maximum depth, and the dorsal angulation measuring approximately 148°. Head and opercular bones and large postclavicular plates externally ornamented with fine, closely arranged rugæ of enamel. Fin-rays stout and smooth; pectoral fins scarcely twice as large as the pelvic pair, and the latter arising in advance of the middle point of the trunk; dorsal fin consisting of at least 25 rays. Pectinations of the scales delicate and confined to their hinder margin, but conspicuous in all regions of the trunk.

Description of Specimens.—The type specimen is so imperfectly shown, with the caudal region so erroneously restored, in Egerton's original figure that it is re-drawn in Pl. XVII, fig. 1. The head-bones are much crushed and broken, but several can be recognised; the ventral region of the trunk is slightly deepened by crushing; and the caudal region is only vaguely indicated in impression, with a few displaced remains of the fulcra of the anal fin. The characters of the species are better seen in a fine though distorted specimen (Pl. XVII, fig. 2) described by Mansel-Pleydell (*loc. cit.*, 1889), and various features are shown by several more imperfect specimens in the British Museum and the Dorset County Museum.

The deep laterally compressed skull is remarkable for the great development of the slender ethmoidal region, which is as long as the frontals. The parietals (Pl. XVII, fig. 3, *pa.*) are conspicuous behind, each slightly longer than broad, and covered with the rugose enamelled ornament, which tends to an antero-posterior direction in the anterior half but is disturbed by a large transverse slime-canal in the posterior half. The squamosals have not been clearly observed. The frontals (fig. 3, *fr.*) are about twice as long as the parietals, slightly arched over the large orbit and not much tapering in front. In their postorbital expansion the rugose ornament is sparse, but in their interorbital portion it is more conspicuous and tends towards longitudinal ridges. The frontals are also traversed by a pair of very large slime-canals. No superficial plates have been observed on the long ethmoidal region. The stout parasphenoid is usually distinct crossing the orbit. The cheek seems to have been covered with plates, but only those in the upper part of the circumorbital ring have hitherto been clearly observed (fig. 3, *co.*). They are narrow, four in number, the foremost the longest and tapering in front, all finely ornamented with irregular tubercles of ganoine.

The mandibular suspensorium is much inclined forwards, so that the quadrate articulation is beneath the front border of the orbit and the mouth is very small. One of the bones of the pterygo-palatine arcade, either palatine or ectopterygoid (Pl. XVII, fig. 4), bears a close series of comparatively large and stout conical teeth. The marginal teeth of both jaws, also in close series, are slender and styliform. The maxilla, crushed and broken in the type specimen (Pl. XVII, fig. 1, *mx.*), is a rather large smooth lamina of bone, deepest behind, and not showing any teeth. A fragment in another specimen (B.M. no. P. 3614), however, seems to bear a very small styliform tooth. The dentary (*d.*) is comparatively slender, and marked by large perforations for the slime-canal.

The opercular apparatus is complete. The preoperculum (Pl. XVII, figs. 1, 2, *pop.*), which is widely expanded at the angle and has a relatively small lower limb, is smooth except at the upper end, where it bears a few oblique ridges. It is deeply excavated with a few large pits for the well-developed slime-canal. The operculum (*op.*), which is about two-thirds as wide as deep, is closely ornamented, except at its upper end, with coarse tubercles of ganoine which tend to fuse into short ridges radiating from the point of suspension. The suboperculum (*sop.*), which is about one-quarter as deep as the operculum, is similarly ornamented, and bears a large smooth ascending process in front. The triangular interoperculum (*iop.*), which is about as deep as broad, and the large upper branchiostegal rays (*br.*) are also ornamented in their lower portion by tubercles of ganoine which are more or less fused into short oblique ridges. The extent of the branchiostegal apparatus between the mandibular rami is unknown.

The vertebral axis is usually obscured by the dense squamation, but broad well-ossified vertebral rings have been seen in one fragmentary specimen (B.M. no. P. 3614).

The occipital border of the cranium is overlapped by supratemporals (fig. 1, *st.*), of which it can only be stated that they bear a row of relatively large pits for the transverse slime-canal. A large triangular post-temporal (fig. 1, *ptt.*) on each side, as usual, supports the pectoral arch; it is ornamented with tubercles and short ridges of ganoine only at its hinder border. The supraclavicle (fig. 1, *scl.*) is ornamented only at its upper end and above the passage of the slime-canal, and tapers to a blunt point below. The clavicle (fig. 1, *cl.*) is relatively large and smooth, sometimes with traces of tubercles of ganoine on its posterior angle. There are the usual three large postclavicular plates (fig. 2, *pcl.*) ornamented like the opercular bones. The pectoral fin in the type specimen measures 2·5 cm. in length and comprises nine or ten rays, of which the foremost is the stoutest and fringed with small fulcra; the basal half of each ray is undivided, but the distal half is closely articulated and more or less subdivided. The pelvic fin is smaller, with very large uniserial fulcra. The dorsal fin (Pl. XVII, fig. 2) comprises about twenty-five well spaced stout rays, which are closely articulated and subdivided for the greater part of their length. Its foremost rays are ornamented with smooth longitudinal ridges of ganoine, and its anterior border is fringed with large uniserial fulcra, of which a few are basal and gradually increase in length. The anal fin is very small, with about four rays. The caudal fin is clearly forked, with about eight enamelled rays in each lobe, and the fulcra on the upper lobe larger than those on the lower lobe.

All the scales, except a few between the base of the pectoral fins, are completely covered with smooth enamel, and most of them even in the caudal region are finely pectinated at the hinder border. The total number of transverse series, counted along the course of the lateral line, is about 40; and the number of scales in a transverse series above the pelvic fins is about 12. The principal scales of the flank in the abdominal region, which have often a slightly convex pectinated edge, are from two to three times as deep as broad, while those of the flank in the caudal region are also deeper than broad. Nearly all the scales dorsally and ventrally are at least as deep as broad, and very few are destitute of posterior serration or pectination, though their shape is often more or less irregular. The scales immediately bordering the dorsal fin are especially peculiar in shape, truncated and usually widest at their upper end, concave at their posterior pectinated edge; four small smooth-edged scales of irregular shape are separated from them beneath the fulcra and foremost ray of this fin. Three or four smooth-edged diamond-shaped ridge-scales, not much enlarged, occur on the caudal pedicle between the dorsal and caudal fins. As shown in the type specimen, a few smooth-edged scales also occur on the narrow ventral face of the abdominal region; while some of the smaller scales between and in front of the base of the pectoral fins bear only isolated ridges and tubercles of enamel. The lateral line traverses the seventh row of scales above the ventral border in the abdominal region. Externally it is only feebly marked by a ridge and notch on each scale, with an occasional perforation, but on the inner face it forms a deep groove (Pl. XVII, fig. 5, *l.*). The smaller perforations of an upper slime-canal are also seen on the scales at the base of the dorsal fin. Nearly all the scales are strengthened by a broad median vertical ridge on the inner face, and most are united by a broad peg-and-socket articulation (Pl. XVII, fig. 5).

Horizon and Localities.—Middle Purbeck Beds: Swanage, Dorset; Tisbury, Wiltshire.

Genus ENCHELYOLEPIS, novum.

Generic Characters.—Head large, snout acute; trunk gradually tapering from the occiput backwards. Marginal teeth much elongated, closely arranged. Notochord invested with delicate ring-vertebræ; neural and hæmal arches especially short and stout. Fins consisting of robust, bifurcating rays, without fulcra except in the caudal; pectoral fins larger than the pelvic pair; dorsal fin with very stout supports, arising immediately behind the occiput and extending continuously to the caudal pedicle; anal fin small; caudal fin not forked. Scales very thin and deeply overlapping, the exposed portion rounded and exhibiting a regularly reticulated structure; no enlarged ridge-scales on caudal pedicle.

Type Species.—*Enchelyolepis andrewsi*, from the English Purbeck Beds.

Remarks.—This genus is known only by two small specimens representing two species, the one from the Purbeck Beds described below, the second from the Upper Portlandian of Savonnières-en-Perthois, Meuse, France. The latter was described under the name of *Macrosemius pectoralis* by H. E. Sauvage (Bull. Soc. Géol. France [3] vol. xi, 1883, p. 477, pl. xii, fig. 17), and is re-figured for comparison with the Purbeckian species in Pl. XVII, fig. 7. The genus is distinguished from *Macrosemius* and all other Macrosemiidæ by its peculiar thin squamation; it also differs from all genera of this family, except *Petalopteryx*, in the stoutness of its neural and hæmal arches and of the dorsal and anal fin-supports.

The scales of *Enchelyolepis* are perhaps most closely similar to those of *Amia*, but the reticulate structure of their exposed portion is rather suggestive of that of the scales of eels.

1. Enchelyolepis andrewsi, A. S. Woodward. Plate XVII, fig. 6.

1895. *Macrosemius andrewsi*, A. S. Woodward, Geol. Mag. [4] vol. ii, p. 148, pl. vii, fig. 3; and Catal. Foss. Fishes B. M., pt. iii, p. 180.

Type.—Nearly complete fish; British Museum.

Specific Characters.—A very small species about 35 mm. in length. Length of head with opercular apparatus considerably exceeding its maximum depth, and contained about three and a half times in the total length to the base of the caudal fin; maximum depth of trunk twice that of caudal pedicle. Pelvic fins inserted about midway between the pectoral and caudal fins; dorsal fin much less deep than the trunk, with about 25 slender rays; anal fin with 7 or 8 rays. All scales apparently broader than deep.

Description of Specimen.—The type specimen, discovered by the Rev. W. R. Andrews, is shown somewhat enlarged in Pl. XVII, fig. 6, and is preserved in counterpart. The remains of the head are merely sufficient to show that it is shaped as in the Macrosemiidæ, with a comparatively small terminal mouth and a close series of styliform teeth in both jaws. There are apparently pro-otic and opisthotic ossifications in the lateral wall of the brain-case. An epihyal and a relatively large ceratohyal bear a few branchiostegal rays, of which four are slender and much curved. In the vertebral axis there are about 35 segments, of which 14 may be reckoned as abdominal. Centra seem to be represented by delicate complete cylinders in the abdominal region, but by not more than small pleurocentra and hypocentra in the caudal region—a condition possibly due to the immaturity of the specimen. In *Enchelyolepis pectoralis*, however, complete vertebral rings are seen throughout the axis (Pl. XVII, fig. 7). The ribs are comparatively slender, and are far from reaching the ventral border of the fish. The neural arches generally, and the hæmal arches in the caudal region, are very short and stout and much inclined to overlap. A few of the anterior neural arches are expanded at the upper end, which is seen to be forked in *E. pectoralis*; about four of the neural arches at the base of the caudal fin are longer than the others and very closely arranged. Eight or nine elongated hæmals are supports of the caudal fin. The stout clavicle is considerably expanded above the pectoral fin, which is inserted close to the ventral border of the fish, but is only fragmentary in the fossil. In *E. pectoralis* the pectoral fin consists of about 12 articulated rays supported by 4 well-calcified basals which rapidly decrease in size from below upwards. Each pelvic fin-support is characterised by its relatively large and wide proximal triangular expansion; while between this support and the 5 or 6 pelvic fin-rays in *E. pectoralis* there seem to be small nodular baseosts. The dorsal fin seems to have been equally elevated throughout its length, and some of the rays clearly show their distal bifurcation. All its 25 supports (fig. 6 *a*) are especially stout, larger than the neural spines, and sharply curved forwards at their pointed lower end. Between each of these supports and its corresponding ray is intercalated a very short bony rod, as in *Amia*[1]—an arrangement seen again in *Enchelyolepis pectoralis*, where there is also another small nodule of bone between the short rod and the fin-ray (fig. 7 *a*). The anal fin, with its 7 rays, exhibits similar stout supports. Of the caudal fin only the base is preserved; but in *E. pectoralis* it is nearly complete and shows the rounded or truncated posterior border. Traces of scales are vaguely seen over nearly the whole of the trunk in the fossil, but they must have been extremely thin. They appear to be at least as long as deep and much overlapping, the large covered portion being marked only by the usual concentric lines of growth, while the small exposed portion, which is rounded at the hinder border, has a regular reticulate structure (Pl. XVII, fig. 6 *b*).

[1] T. W. Bridge, in 'The Cambridge Natural History,' vol. vii (1904), p. 235, fig. 136.

Traces of similar scales also occur in *E. pectoralis*, some fragments overlying the pelvic bones being especially clear (Pl. XVII, fig. 7 b). The only fulcral scales are at the base of the upper lobe of the tail.

Horizon and Locality.—Middle Purbeck Beds: Vale of Wardour, Wiltshire.

Family EUGNATHIDÆ.

Genus **CATURUS**, Agassiz.

Caturus, L. Agassiz, Neues Jahrb. f. Min., etc., 1834, p. 387.
Uræus, L. Agassiz, Poiss. Foss., vol. ii, pt. i, 1833, p. 12, (*non Uræus*, Wagler, 1830).
Conodus. L. Agassiz, tom. cit., pt. ii, 1844, p. 105 (name only).
Strobilodus, A. Wagner, Abhandl. k. bay. Akad. Wiss., math.-phys. Cl., vol. vi, 1851, p. 75.

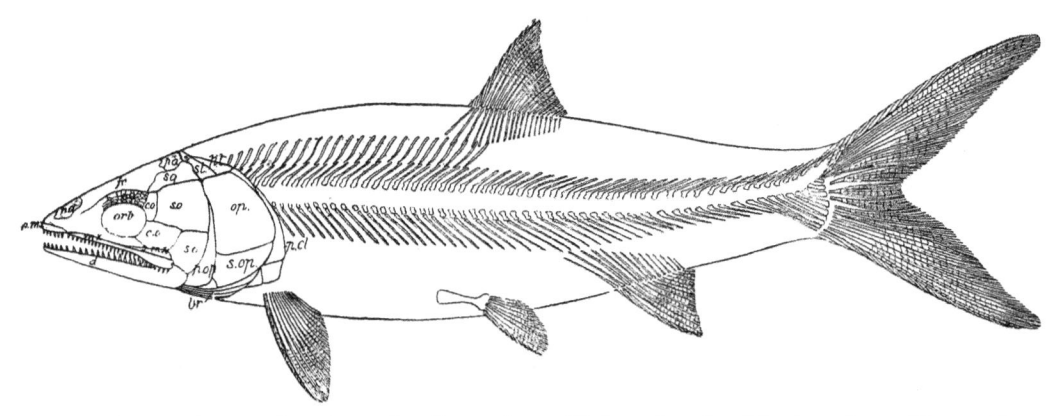

FIG. 26.—*Caturus furcatus*, Agassiz; restoration, scales omitted, much reduced in size.—Lower Kimmeridgian (Lithographic Stone); Bavaria. *br.*, branchiostegal rays; *co.*, circumorbitals; *d.*, dentary; *fr.*, frontal; *mx.*, maxilla; *na.*, nasal; *op.*, operculum; *orb.*, orbit; *pa.*, parietal; *pcl.*, postclavicular plates; *pmx.*, premaxilla; *pop.*, preoperculum; *pt.*, post-temporal; *smx.*, supramaxilla; *so.*, postorbitals; *sop.*, suboperculum; *sq.*, squamosal; *st.*, supratemporal.

Endactis, P. M. G. Egerton, Figs. and Descripts. Brit. Organic Remains (Mem. Geol. Surv.), dec. ix, 1858, no. 4.
Thlattodus, R. Owen, Geol. Mag., vol. iii, 1866, p. 55.
Ditaxiodus, R. Owen, tom. cit., 1866, p. 107.

Generic Characters.—Trunk elongate-fusiform. External head-bones and opercular bones feebly ornamented with rugæ and tuberculations, all except the cheek-plates robust; upper circumorbitals subdivided into small tesseræ; snout obtusely pointed, and maxilla straight or with a somewhat concavely arched dentigerous border; teeth relatively large and tipped with enamel, arranged in a sparse series on the margin of the jaws, smaller on the palatine and on the splenial, where they are in single series anteriorly, minute and almost granular on the other inner bones; preoperculum narrow, nearly smooth; operculum deep, much broader below than above, and suboperculum of moderate size. Ossifications round the notochord insignificant or absent in the smaller species, consisting only of separate hypocentra and pleurocentra in the largest species; ossified ribs slender, not reaching the ventral border of the abdominal region. Fulcra biserial, well-developed on all the fins, those of the pectoral being especially elongated and sometimes in part fused together. Pectoral much exceeding the pelvic fins in size, but the latter well-developed; dorsal and anal fins triangular in shape, the former arising opposite or immediately behind the pelvic fins; caudal fin deeply forked. Scales thin, smooth, feebly crimped or in part tuberculated, deeply overlapping, and none much deeper than broad; a few anterior series quadrangular and sometimes united with peg-and-socket, the others more or less cycloidal, and very few narrowed near the ventral border. Lateral line inconspicuous.

Type Species.—*Caturus furcatus* (L. Agassiz, Poiss. Foss., vol. ii, pt. ii, 1842—44, p. 116, pl. lvi a) from the Lithographic Stone (Lower Kimmeridgian) of Bavaria. See Text-fig. 26.

Remarks.—This genus is represented only by fragments in the Wealden and Purbeck Formations. It is known by many nearly complete fishes from the

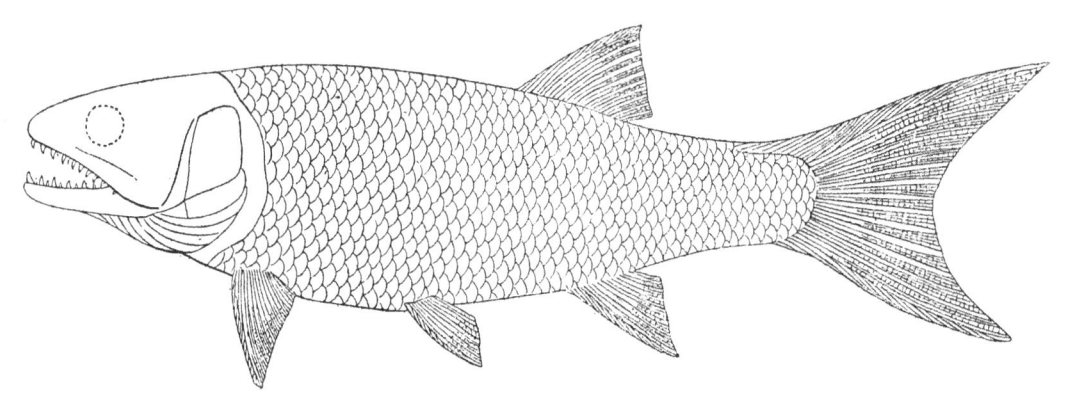

FIG. 27.—*Callopterus insignis*, Traquair; restoration, showing scales, much reduced in size.—Wealden; Bernissart, Belgium. After R. H. Traquair.

Lithographic Stone of Germany and France; and its cranial osteology is well displayed in numerous fragmentary specimens discovered by Mr. Alfred N. Leeds in the Oxford Clay near Peterborough (A. S. Woodward, Ann. Mag. Nat. Hist. [6] vol. xix, 1897, pp. 292—297, pls. viii, ix).

1. **Caturus (Callopterus?) latidens**, sp. nov. Text-figure. 28.

Type.—Imperfect skull; British Museum.

Specific Characters.—Marginal teeth broad, laterally compressed near the apex, and bluntly pointed; those of middle of maxilla about as deep as the bone at their insertion.

Description of Specimen.—The type and only known specimen was discovered by Mr. S. H. Beckles in a waterworn fragment of Wealden ironstone on the beach near Hastings. It exhibits only remains of the head, which are very fragmentary (Text-fig. 28).

84 WEALDEN AND PURBECK FOSSIL FISHES.

The skull must have measured originally about 10 cm. in length, but the parietal and occipital regions are missing. The remains of the thick rugose frontal bones (*fr.*) show the usual deep interdigitation of their median suture in the region between the postfrontals (sphenotics). The tapering anterior end of the narrow squamosal is also seen bordering the hinder end of the frontal on the left side (*sq.*). The cheek-plates are marked with a finer rugosity than that of the cranial roof, with some tuberculation. Both the postorbitals (*so.*) are imperfect, but the lower seems to be the larger, its subdivision not being clear. The postero-

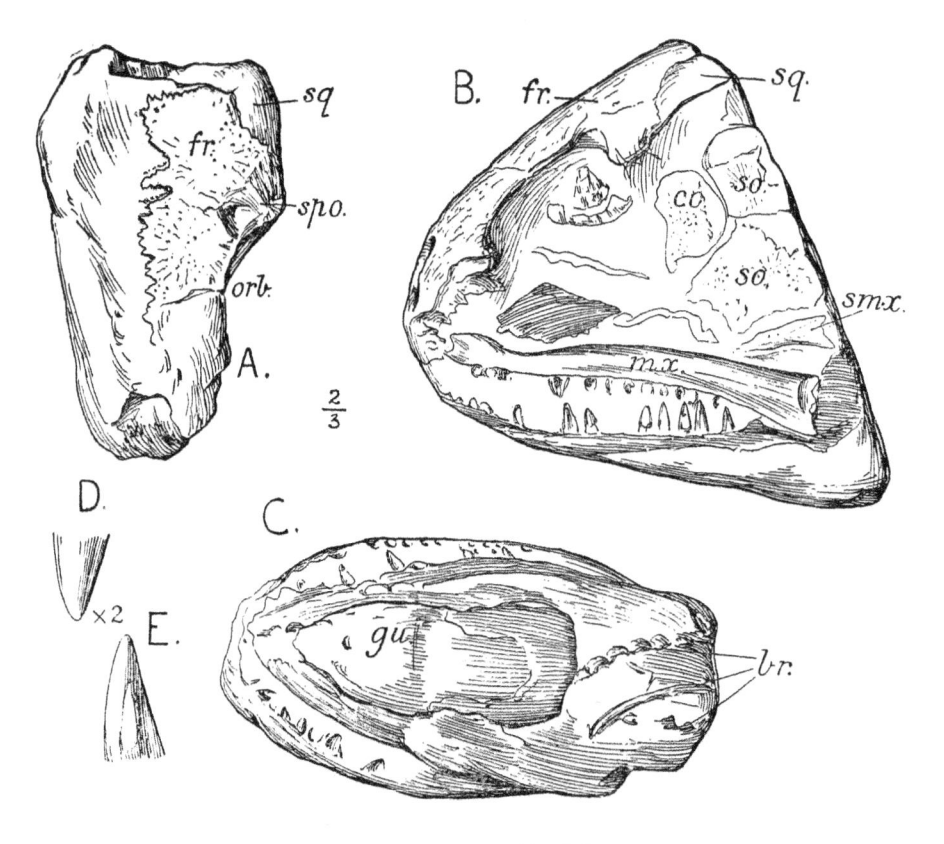

Fig. 28.—*Caturus latidens*, sp. nov.; imperfect head, upper (A), left side (B), and lower (C) views, two-thirds nat. size, with upper (D) and lower (E) teeth enlarged twice.—Wealden; Hastings, Sussex. Beckles Collection (B. M. no. P. 6360). *br.*, branchiostegal rays; *co.*, posterior circumorbital; *fr.*, frontal; *gu.*, gular; *mx.*, maxilla; *orb.*, orbit; *smx.*, supramaxilla; *so.*, postorbitals; *spo.*, sphenotic (postfrontal); *sq.*, squamosal.

superior circumorbital (*co.*), as usual, is deeper than wide, and there seem to be remains of the irregular superior circumorbitals. One of the antorbitals, a little displaced, is an elongated irregularly rhomboidal plate. The maxilla (*mx.*) exhibits the ordinary constriction near its anterior end, and is not much deepened behind; its outer face is comparatively smooth, marked only by a few irregular longitudinal grooves. Its hinder end was originally bordered by a single narrow supramaxilla, similarly ornamented, which tapers to a point in front (*smx.*). Most of the maxillary teeth are broken away, exposing their very large pulp-cavity.

They are nearly round in section at their base, but, as shown in the right maxilla, their blunt apex is laterally compressed (Text-fig. 28, D). A cluster of nearly similar teeth, curved and apparently less compressed at the apex, occurs in front of the right maxilla on a thick piece of bone which may represent the palatine. The mandible is very fragmentary, but its lower border must have been much curved inwards, and its coronoid region rises abruptly into the usual considerable elevation. Its teeth (E) resemble those of the maxilla (D), but seem to be somewhat larger. The gular plate (*gu.*), which occurs in position between the mandibular rami, is especially large, reaching as far backwards as the hinder end of the tooth-bearing border of the dentary. Traces of the anterior branchiostegal rays (*br.*) show that they are thick rather than laminar.

Remarks.—As suggested by the shape of the teeth, it is possible that this species may not belong to *Caturus*, but to the closely allied genus *Callopterus* (with a more remote dorsal-fin), which has already been recorded from the Wealden of Belgium by R. H. Traquair, 'Les Poissons Wealdiens de Bernissart' (Mem. Mus. Roy. Hist. Nat. Belg., vol. v, 1911, p. 34, pl. vi). See Text-fig. 27, p. 83.

Horizon and Locality.—Wealden : Hastings, Sussex.

2. **Caturus purbeckensis**, A. S. Woodward. Plate XIX, figs. 1, 2.

1890. *Strobilodus purbeckensis*, A. S. Woodward, Proc. Zool. Soc., p. 350, pl. xxix, fig. 4.
1895. *Caturus (Strobilodus) purbeckensis*, A. S. Woodward, Catal. Foss. Fishes Brit. Mus., pt. iii, p. 348.

Type.—Head; British Museum.

Specific Characters.—Head with opercular apparatus usually attaining a length of from 10 to 15 cm.; external bones without ornament. Maxilla a little curved downwards behind, where its teeth become comparatively small and slender; teeth of middle of maxilla much deeper than the bone at their insertion. Mandible very slender, pointed, and tending to curve upwards in front; height of middle dentary teeth nearly equal to depth of the bone at their insertion. All the teeth tumid, often with an external indent, at their base, becoming very slender in their incurved apical half, and tipped by a laterally compressed cap of translucent ganodentine with prominent edges.

Description of Specimens.—This species is known only by the imperfect specimen of the head and adjoining parts shown in Pl. XIX, fig. 1, and by detached jaws.

The cranium is so much crushed in the type specimen (Pl. XIX, fig. 1) that its characters can only be vaguely seen. The roof-bones are marked by a faint rugosity and numerous irregular fine pittings. The postfrontal (sphenotic) is well ossified; and the stout parasphenoid, where it crosses the orbit and eventually underlaps the vomers, is a depressed lamina. The postorbital cheek-plates and the large hinder circumorbitals are smooth, exhibiting only very fine scattered pittings or punctations. The upper circumorbitals, as usual in *Caturus*, seem to have been irregularly subdivided. The mandibular suspensorium is inclined backwards, and the stout hyomandibular is visible beneath the cheek-plates. The maxilla (*mx.*) is smooth, with slight longitudinal grooving, and its hinder end is overlapped by a long and narrow supramaxilla (*smx.*). It exhibits its usual slight sinuosity, and is a little deepened where its oral border curves downwards behind. The relatively large teeth with their nearly square base fused to the bone, their extensive inner cavity, and their slender incurved apical half, with a triangular tip of ganodentine, are well seen: they become especially small and slender behind. The extended premaxilla (*pmx.*), with five or six tooth-sockets, seems to have borne rather larger teeth. The mandible, shown both in the type specimen and in a smaller specimen (Pl. XIX, fig. 2), displays its characteristic upturned pointed symphysial end, and rises behind into a short coronoid region, in which the ordinary separate coronoid bone (*co.*) and small angular bone (*ag.*) can be distinguished. The dentary (*d.*) is a nearly smooth bone, punctate in part and marked by a row of pits along the course of the slime-canal; the angular bone is strongly punctate. The teeth resemble those of the maxilla, but are somewhat larger. The teeth of both jaws often exhibit a median indent on their outer face at the base; and the bone round their insertion is sometimes marked with very fine short radiating grooves or crimpings.

A dentary bone in the Sedgwick Museum, Cambridge, is one-third larger than that of the type specimen; and part of another dentary in the Beckles Collection (B.M. no. P. 6388) is equally large.

The remains of the opercular bones in the type specimen are smooth, apart from fine punctations; and the clavicle (*cl.*) is only marked by a few vertical lines within its overlapped margin. Ossified hypocentra (*hy.*) are seen in the anterior part of the trunk, besides neural arches and ribs, partly obscured by the usual thin scales. The base of the pectoral fin (*pct.*) shows the stoutness of its smooth, closely adpressed rays.

Horizon and Locality.—Middle Purbeck Beds: Swanage, Dorset.

3. **Caturus tenuidens**, A. S. Woodward. Plate XIX, figs. 3, 4.

1895. *Caturus tenuidens*, A. S. Woodward, Geol. Mag. [4], vol. ii, p. 151, pl. vii, figs. 7, 8.

Type.—Mandibular ramus; British Museum.

Specific Characters.—Teeth very slender, usually incurved at the apex, less swollen at the base than in the type species and in *C. purbeckensis*, usually well-spaced in the jaw. Dentary bone almost smooth, curved a little upwards at its pointed symphysial end; height of teeth in middle of dentary series much less than the depth of the bone at their insertion.

Description of Specimens.—This species is still known only by isolated dentary bones and maxillæ, which are not uncommon in the Middle Purbeck Beds of Swanage. As they are smaller than some of the corresponding bones of *C. purbeckensis*, they might at first be regarded as representing the young of the latter species; but if dentaries of nearly the same size be compared (as in Pl. XIX, figs. 2, 4), the more slender proportions of *C. tenuidens* are evident. All the teeth, when well preserved, exhibit the usual little triangular apex of translucent ganodentine. The characteristic maxilla shown in inner view in Pl. XIX, fig. 3, is nearly complete, displaying the shape of the hinder slight expansion and the depressed anterior half of the bone.

Horizon and Locality.—Middle Purbeck Beds: Swanage, Dorset.

Genus NEORHOMBOLEPIS, A. S. Woodward.

Neorhombolepis, A. S. Woodward, Proc. Geol. Assoc., vol. x, 1888, p. 304.

Generic Characters.—Trunk elongate-fusiform, more or less laterally compressed, and head relatively large. External head-bones and the opercular bones stout, more or less ornamented with tubercles and rugæ of ganoine, but no prominent bosses or outgrowths. Maxilla with a straight tooth-bearing border and a long supramaxilla; teeth conical, in regular series, large and hollow on the margin of the jaw, not in sockets; a patch of minute teeth on the parasphenoid. Suboperculum at least half as large as the operculum, which is quadrangular but truncated at the postero-superior angle. Vertebral centra either ring-shaped or completely ossified. Fulcra well-developed on the pectoral fins, probably on the other fins. Scales rhombic and thick, with a wide overlapped margin not produced at the angles, and the peg-and-socket articulation feeble or wanting; superficial ganoine nearly smooth; few principal flank-scales as deep as broad, the majority broader than deep, and those of numerous ventral series at least twice as broad as deep, sometimes subdivided.

Type Species.—*Neorhombolepis excelsus* (A. S. Woodward, Proc. Geol. Assoc., vol. x, 1888, p. 304, pl. i, fig. 1; also Foss. Fishes of English Chalk—Pal. Soc., 1909—p. 158, pl. xxxiv, fig. 1) from the English Lower Chalk.

Remarks.—A detailed study of the type specimen of the Wealden species described below, modifies the original definition of *Neorhombolepis* by making known the fulcra on the pectoral fins.

1. Neorhombolepis valdensis, A. S. Woodward. Plate XVIII.

1895. *Neorhombolepis valdensis*, A. S. Woodward, Catal. Foss. Fishes B.M., pt. iii, p. 356, pl. viii, fig. 5.

Type Specimen.—Imperfect fish; British Museum.

Specific Characters.—As large as the type species, skull attaining a length of about 8 cm. External head-bones ornamented with large flattened tubercles of ganoine. Scales smooth but often marked with a few fine punctations; hinder border of all the principal abdominal scales delicately serrated.

Description of Specimen.—The type and only known specimen was discovered by Mr. S. H. Beckles in a waterworn fragment of Wealden ironstone on the beach near Hastings. The fish is curled up, dislocated across the end of the abdominal region, and partly exposed on both sides of the pebble. The remains are shown of the natural size in Pl. XVIII, figs. 1—4.

The roof of the skull must have been nearly flat, with the postorbital region about two-thirds as long as broad, and the interorbital region much excavated by the relatively large orbits. Of the roof-bones only portions of the parietals (*pa.*) and squamosals (*sq.*) are well seen, these evidently uniting with the frontals in a deeply interdigitating suture. They have a somewhat wrinkled surface, irregularly ornamented with large flattened tubercles of ganoine. The parasphenoid (fig. 2, *pas.*), seen on the left side of the fossil, extends as far back as the occiput. Its hinder portion is laterally compressed to a sharp median ridge below; between its lateral wings the lower face widens to bear a small elongate-oval patch of minute teeth; and further forwards it evidently becomes more expanded. Its lateral wing is relatively large and bifurcated, the anterior limb rising as usual to meet the postfrontal or sphenotic (*ptf.*), while the posterior limb reaches one of the hinder otic elements. Remains of cheek-plates occur on the right side, ornamented like the bones of the cranial roof with large, irregular, flattened tubercles of ganoine. The postorbitals (*po.*) must have been relatively large; and there are small circumorbitals (*co.*) above the eye.

The right hyomandibular (*hm.*) occurs in its natural position and shows that the mandibular suspensorium is nearly vertical. This bone is laterally compressed and more expanded above than below, bearing a large and deep prominence for the suspension of the operculum. At least half of its lower end articulates with the symplectic (*sy.*), which is widest above, then becomes much constricted in its lower half, and thickens again a little at its lower end, which may have articulated with the mandible as in *Amia* and *Caturus*. The quadrate (*qu.*) and pterygoid bones are thin laminæ, only imperfectly shown on the right side of the fossil. Traces of a patch of small teeth are recognisable below the anterior end of the ectopterygoid. A fragment and a partial impression of the long and slender maxilla (*mx.*) show a regular series of stout, hollow, conical teeth. The hinder end of the mandible (*md.*), with part of the elevated coronoid region, is seen on the right side; and a splintered fragment of the dentary, with its regular series of stout conical teeth, occurs on the left (fig. 4).

The opercular apparatus is almost unknown; but one bone on the right side (fig. 1, *x.*) may be the interoperculum, while traces of broad branchiostegal rays are seen on the left (fig. 2, *br.*). The supposed interoperculum, which is exposed from the inner face, is about twice as long as its greatest depth.

Traces of ossified centra, perhaps only cylinders, are seen throughout the vertebral column; and in the middle of the fish, where the column is dislocated, they are well displayed in side view (*v.*). They are smooth, mostly as long as deep, not mesially constricted, but laterally compressed above and below to a longitudinal ridge for the support of the arch. Some of the neural arches are seen to be not completely fused with the centra; and two of those in the anterior part of the caudal region exhibit the relatively small and slender neural spine (*n. a.*) bent sharply backwards from the deep pedicle.

In the pectoral arch the supraclavicle (fig. 1, *scl.*) is relatively large, three times as deep as wide, truncated and hollowed at the upper articular end, bluntly rounded below. Its exposed portion bears large flattened tubercles of ganoine, which are partly fused together. The slender clavicle, as far as seen on the left side (fig. 2, *cl.*), is not enamelled. The rather large postclavicular scales (figs. 1, 2, *pcl.*), so far as preserved, are completely covered with ganoine except at the overlapped margin, and are partly ornamented with coarse flattened rugæ and tuberculations. They are evidently arranged as in the typical *Eugnathus*, the uppermost being the largest, deep and triangular, its apex extending upwards somewhat above the lower end of the supraclavicle. As shown on the left side, the pectoral fin (fig. 3, *pct.*) comprises a few more than 20 rays, of which each is undivided in its proximal two-thirds but becomes very finely branched and articulated distally. The foremost ray is enlarged at its proximal end where it projects upwards above the others; and it is fringed with elongated, deeply overlapping fulcra, of which the two uppermost are the stoutest and must have been in direct contact with the basal supports. Both the fulcra and some of the anterior fin-rays bear traces of ganoine. On the right side of the fossil, only two fragments of the pectoral fin remain (fig. 1, *pct.*), but the characteristic elongated fulcra are seen fringing the articulated distal portion. Of the other fins merely imperfect traces of the dorsal (*d.*) and caudal (*c.*) are exposed.

The principal scales of the flank and some of the narrow ventral scales in the abdominal region exhibit a very fine sparse pitting of the enamel and a regular delicate serration or pectination of the hinder border, but the other scales are smooth. They are most deeply overlapping, with the best developed peg-and-socket articulation, on the anterior part of the flank. Very few of the flank-scales are as deep as broad, most being longer than deep, and those near the ventral margin especially elongated. Many of the ventral scales in the abdominal region are subdivided into small irregularly rhombic scales which only slightly overlap (fig. 3). Some of the scales of the lateral line are pierced by a simple short vertical slit.

Horizon and Locality.—Wealden : Hastings, Sussex.

Family AMIIDÆ.

Genus **AMIOPSIS**, Kner.

Amiopsis, R. Kner, Sitzungsb. k. Akad. Wiss. Wien, math.-naturw. Cl., vol. xlviii, pt. i, 1863, p. 126.

Generic Characters.—Trunk elongate and laterally compressed. Head large; all the marginal teeth large and conical, but those of the dentary largest; inner teeth smaller; maxilla laterally compressed and deepened behind. Vertebral centra completely ossified in the adult, biconcave, the hypocentra and pleurocentra forming distinct alternating discs in the caudal region; each centrum impressed with three or more extended pits on its side; ribs short. Fins without fringing fulcra, the rays articulated and branching; dorsal fin occupying not more than one-third of the back; anal fin small and short-based; caudal fin with convex

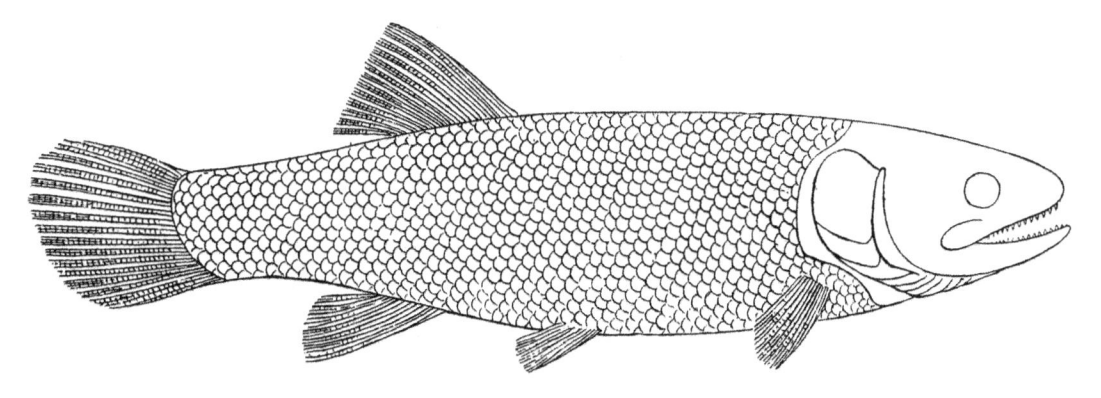

FIG. 29.—*Amiopsis dolloi*, Traquair; restoration, showing scales, much reduced in size.—Wealden; Bernissart, Belgium. After R. H. Traquair.

hinder border. Scales almost oval in shape, the long axis horizontal; exposed portion thickened.

Type Species.—*Amiopsis prisca* (R. Kner, *loc. cit.*, 1863, p, 126, pl. i), from the Cretaceous of Mrzlek, valley of the Isonzo, Istria.

Remarks.—This genus was not satisfactorily defined until 1895, when Kramberger described well-preserved specimens of the typical species (Djela Jugoslav. Akad., vol. xvi, p. 12, pl. iii, fig. 2, pl. iv). The pittings in the side of the vertebral centra distinguish it from *Megalurus* (L. Agassiz, Poiss. Foss., vol. ii, pt. i, 1833, p. 13), which occurs typically in the Lithographic Stone (Lower Kimmeridgian) of Germany, and has also been found in the Lower Cretaceous of Bahia, Brazil (*Megalurus mawsoni,* A. S. Woodward, Ann. Mag. Nat. Hist. [7] vol. ix, 1902, p. 87, pl. ii). Two species, *Amiopsis dolloi* (Text-figs. 29, 30) and *A. lata,* from the Wealden of Bernissart, Belgium, have been described by R. H. Traquair, Mém. Mus. Roy. d'Hist. Nat. Belg., vol. v, 1911, pp. 37, 42, pls. vii, viii.

1. **Amiopsis damoni** (Egerton). Plate XIX, figs. 5, 6.

1858. *Megalurus damoni*, P. M. G. Egerton, Figs. and Descript. Brit. Organic Remains (Mem. Geol. Surv.), dec. ix, no. 8, pl. viii.
(?) 1873. *Megalurus damoni*, V. Thiollière, Poiss. Foss. Bugey, pt. ii, p. 22, pl. ix.
1895. *Megalurus damoni*, A. S. Woodward, Catal. Foss. Fishes, B. M., pt. iii, p. 366.

Type Specimen.—Imperfect fish; British Museum.
Specific Characters.—Attaining a length of about 30 cm. but usually smaller. Length of head with opercular apparatus exceeding maximum depth of trunk and contained about five times in total length of fish; depth of caudal pedicle contained scarcely five times in length from operculum to base of middle caudal fin-rays. External bones slightly rugose, otherwise not ornamented. About 55 vertebræ,

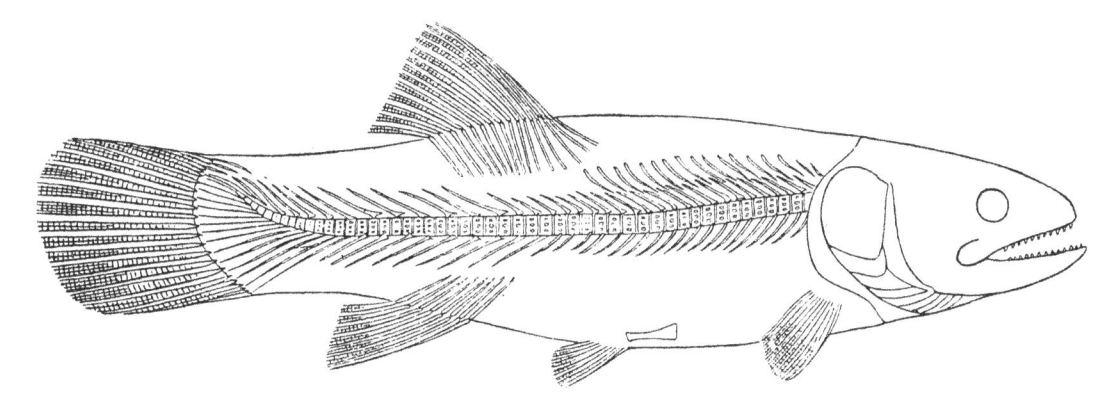

FIG. 30.—*Amiopsis dolloi*, Traquair; restoration, scales omitted, much reduced in size.—Wealden; Bernissart, Belgium. After R. H. Traquair.

half being abdominal. Dorsal fin with about 17 supports, occupying the middle of the back; anal fin with 8 supports, arising opposite the hinder end of the dorsal; pelvic fins inserted opposite the origin of the dorsal. Scales broader than deep, their hinder border not thickened.

Description of Specimens.—The type specimen exhibits most of the generic and specific characters of the fish, and another specimen, even better preserved, is shown of the natural size in Pl. XIX, fig. 5. The bones of the head, though stout, are always much crushed and broken in the fossils. The parietal, squamosal, and frontal bones are as in *Amia*, forming a cranial shield only gently arched from side to side and ornamented with a feeble coarse rugosity. A single pair of supra-temporal plates overlaps the occipital border. The outer margin of each frontal is slightly excavated above the orbit; and in one specimen (B. M. no. 41171) this excavation is occupied by four nearly smooth plates of a narrow circumorbital ring. Traces of similar plates are also seen in the type specimen. The mandibular suspensorium is gently arched forwards so that the quadrate articulation is beneath the hinder border of the orbit. As shown by the type specimen, the maxilla is smooth and shaped as in *Amia*; so far as can be seen in several specimens, the mandible is also Amioid, the dentary bearing a single regular close series of smooth high conical teeth. The preoperculum, as shown in the type specimen, is a narrow arched bone, nearly smooth; the operculum (B. M. no. 41171) must have been nearly as broad as deep; the suboperculum is somewhat more than half as deep as the operculum, with a stout anterior ascending process; the triangular interoperculum is broader than deep; and several of the upper branchiostegal rays are wide laminæ. None of the opercular bones are marked by more than feeble coarse rugæ, which are seen to be radiating on the operculum of the type specimen.

The vertebral centra are well ossified, and all are marked by two principal lateral pits, one above, one below a longitudinal median ridge, which is impressed with a variable number of comparatively small pits. In the abdominal region the centra are about as long as deep, apparently without any processes for the support of the ribs, which are short, stout, and curved. The neural arches in advance of the dorsal fin are surmounted by the usual separate neural spines. In the caudal region the vertebral centra are shorter and deeper, and some clearly alternate with and without neural and hæmal arches in the typical Amioid manner. The neural and hæmal arches are short, slender, and almost symmetrical until the base of the caudal fin, in which about 14 hæmals predominate both in length and in stoutness.

The rays of all the fins are stout and smooth, and all are both closely articulated and divided for the greater part of their length distally (Pl. XIX, fig. 5). A few short basal fulcra, increasing in length, occur at the origin of each fin, but there appear to be no fringing fulcra. Each pectoral fin seems to have comprised 9 or 10 rays, while the pelvic fin has 8 or 9 rays, which are not much more than half as long as the former. The dorsal fin occupies less than the middle third of the back, and the length of its longest rays is somewhat less than the depth of the trunk at their point of insertion. Three or four basal fulcra are distinguishable at its origin. Its 17 supports are all long and stout, expanding at the upper end where they articulate directly with the fin-rays, without the intercalation of any other supports. The anal fin, best shown in the type specimen, is comparatively small, but all its 8 supports are much elongated. The caudal fin, especially well seen in Pl. XIX, fig. 5, is very long and stout and unsymmetrically rounded, as in *Amia*. It is supported by about 14 thickened hæmal arches at the upturned end of the caudal region; and slender basal fulcral scales are conspicuous both above and below.

The deeply-overlapping scales are uniform over the whole of the trunk, longer than deep, and rounded at their free hinder border. The covered portion exhibits only the fine concentric lines of growth, but the exposed portion is thickened and finely pitted with markings which give it an irregularly reticulated appearance. This structure is especially well seen in the type specimen (Pl. XIX, fig. 6), in

Family AMIIDÆ.—Genus **AMIOPSIS**.

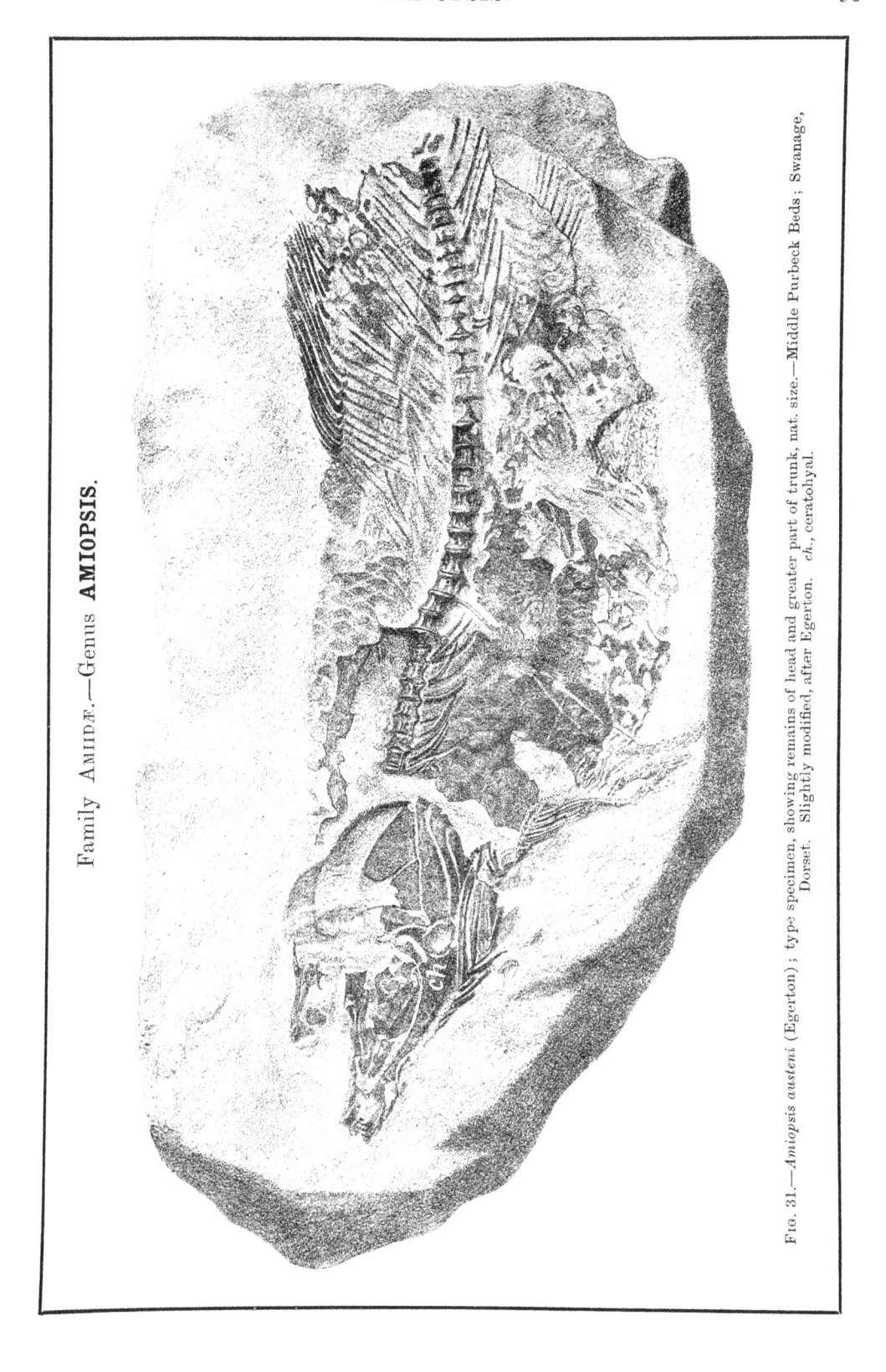

Fig. 31.—*Amiopsis austeni* (Egerton); type specimen, showing remains of head and greater part of trunk, nat. size.—Middle Purbeck Beds; Swanage, Dorset. Slightly modified, after Egerton. *ch.*, ceratohyal.

which some of the anterior scales have the pittings filled with matrix, as drawn somewhat diagrammatically in Egerton's original enlarged figure.

It may be added that in the type specimen some well-fossilised ova are scattered in the abdominal region.

Horizons and Localities.—Lower and Middle Purbeck Beds: Bincombe and Portland, near Weymouth, Dorset.

A larger form of this or another Amioid species is represented by isolated jaws in the Lower Purbeck Beds of Portland. A right maxilla, lacking teeth, is shown in outer view in Pl. XIX, fig. 7; while a left mandibular ramus, lacking the tips of the teeth, is shown from the inner aspect in Pl. XIX, fig. 8. As usual in true Amioids a few of the anterior teeth in the maxilla associated with the latter specimen are stouter than the others.

2. **Amiopsis austeni** (Egerton). Text-figure 31.

1858. *Megalurus austeni*, P. M. G. Egerton, Figs. and Descript. Brit. Organic Remains (Mem. Geol. Surv.), dec. ix, no. 9, pl. ix.
1858. *Attakeopsis (?) austeni*, V. Thiollière, Bull. Soc. Géol. France [2], vol. xv, p. 785.
1895. *Megalurus damoni*, A. S. Woodward, Catal. Foss. Fishes, B. M., pt. iii, p. 366.

Type Specimen.—Imperfect head and abdominal region; British Museum.

Specific Characters.—Imperfectly known, but head smaller, abdominal region deeper, and flank-scales deeper than in *M. damoni*.

Description of Specimen.—This species is still known only by the imperfect and somewhat distorted type specimen (Text-fig. 31). The other fragments, mentioned by Egerton as abundant, are evidently referable to Leptolepidæ. Traces of stout styliform teeth, with a blunt apex, sometimes slightly curved, are seen in both jaws. Remains of the opercular apparatus, much broken, are exposed from within, and a few broad branchiostegal rays occur below them, with the greater part of the supporting epihyal and ceratohyal (*ch.*). The vertebral centra clearly exhibit their lateral pitting, and the short curved ribs are observed to be rather stout. In the anterior part of the caudal region, only alternate centra bear arches. The paired fins are very imperfect, but remains of both occur nearly in their original position, the pectorals being especially large. The dorsal fin has the usual three or four gradually lengthening basal fulcra at its origin. The small anal fin is only partly shown in impression. The comparatively deep scales, with rounded posterior border, are best seen in the upper anterior part of the abdominal region. As in the type specimen of *A. damoni*, there seem to be some fossilised ova.

Horizon and Locality.—Middle Purbeck Beds: Swanage, Dorset.

Family ASPIDORHYNCHIDÆ.

Genus **ASPIDORHYNCHUS**, Agassiz.

Aspidorhynchus, L. Agassiz, Poiss. Foss., vol. ii, pt. i, 1833, p. 14.

Generic Characters.—Rostrum slender, much produced in advance of mandibular symphysis; circumorbital plates very small, postorbitals large; teeth irregular in size, largest on the premaxilla, palatine, and presymphysial bone, reduced to a fine granulation on the inner face of the ectopterygoid; branchiostegal rays short and broad, and gular plate apparently absent. Vertebral centra in form of rings. Fulcra wanting on paired fins, minute on median fins. Pelvic fins situated at about the middle of the trunk; dorsal and anal fins short-based, triangular, remote

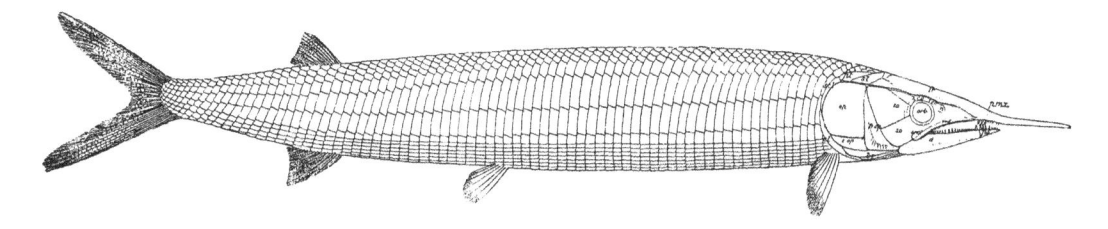

FIG. 32.—*Aspidorhynchus acutirostris* (Blainville); restoration, about one-seventh nat. size.—Lower Kimmeridgian (Lithographic Stone); Bavaria. *br.*, branchiostegal rays; *cl.*, clavicle; *co.*, circumorbitals; *d.*, dentary; *fr.*, frontal; *mx.*, maxilla; *n.*, narial opening; *op.*, operculum; *orb.*, orbit; *pmx.*, premaxilla; *p.op.*, preoperculum; *ps.*, presymphysial; *pt.*, post-temporal; *scl.*, supraclavicle; *smx.*, supramaxilla; *so.*, postorbitals; *s.op.*, suboperculum; *st.*, supratemporal.

and opposed; caudal fin symmetrically forked. Scales robust, smooth or rugose; in two or three deepened series on the flank of the abdominal region, and the foremost scales of the series traversed by the lateral line not deeper than the series below.

Type Species.—*Aspidorhynchus acutirostris* (L. Agassiz, Poiss. Foss., vol. ii, 1833–44, pt. i, p. 14, pt. ii, p. 136, pl. xlvi; *Esox acutirostris*, H. D. de Blainville, Nouv. Dict. d'Hist. Nat., vol. xxvii, 1818, p. 332), from the Lithographic Stone (Lower Kimmeridgian) of Bavaria (Text-fig. 32).

Remarks.—Reis[1] supposed that *Aspidorhynchus* was distinguished from the closely related and partly contemporaneous genus *Belonostomus*, by the intercalation of a supplementary triangular cheek-plate between its upper postorbital and the preoperculum. This determination was adopted both by Zittel[2] and the present writer,[3] but it is now clear that the appearance of such a separate plate is

---

[1] O. M. Reis, 'Ueber *Belonostomus, Aspidorhynchus*, und ihre Beziehungen zum lebenden *Lepidosteus*,' Sitzungsb. k. bayer. Akad. Wiss., math.-phys. Cl., vol. xvii (1887), p. 173, pl. ii, fig. 7.

[2] K. A. von Zittel, Handbuch der Palæontologie, vol. iii (1887), p. 220, fig. 233.

[3] A. S. Woodward, Catal. Foss. Fishes, Brit. Mus., pt. iii (1895), p. 418, text-fig. 42.

due merely to the crushing of the external bones on those beneath.[1] An amended restoration of the type species, *A. acutirostris*, is accordingly given in Text-fig. 32.

Since the last description of *Aspidorhynchus*, two specimens of the skull probably of this genus (but possibly of the closely allied *Belonostomus*), from the Great Oolite of Northampton, have been found to display clearly several of the cranial bones. It is thus interesting to observe the large proportions and median union of the epiotic elements, which have already been described in *Lepidotus* (p. 38). Above the large exoccipitals which meet over the foramen magnum on the occipital face (Text-fig. 33 B, *exo.*), the two epiotics (*epo.*), though slightly obscured by crushing in the matrix, evidently also meet in the middle line. In upper view, however (Text-fig. 33 A), they are clearly seen to be separated forwards

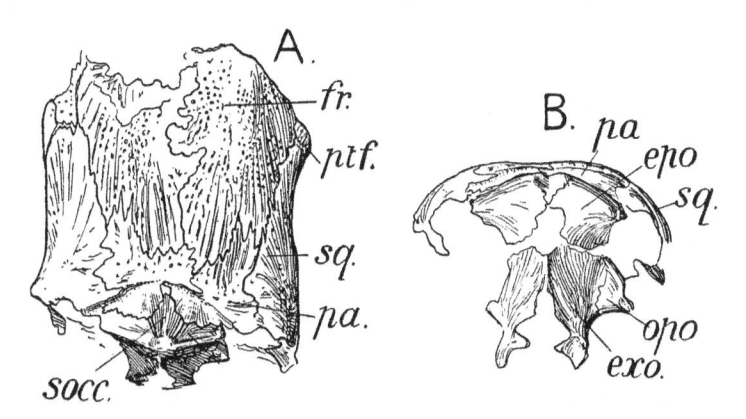

FIG. 33.—*Aspidorhynchus* sp.; imperfect hinder portion of skull, upper view (A) and back view (B), nat. size.—Lower Jurassic (Great Oolite); Kingsthorpe, Northampton. T. Jesson Collection (B. M. no. P. 9843). *epo.*, epiotic; *exo.*, exoccipital; *fr.*, frontal; *opo.*, opisthotic; *pa.*, parietal; *ptf.*, postfrontal (sphenotic); *socc.*, supraoccipital; *sq.*, squamosal. In back view (B) the supraoccipital is not shown, the triangular space between the upper ends of the epiotics being occupied by matrix.

by a small median element of the same texture, which is doubtless an ossified supraoccipital (*socc.*). The very irregular parietals (*pa.*), which overlap these elements, occupy only a small portion of the occipital end of the cranial roof; and their hinder half is depressed and smooth, showing that it was originally covered by the muscles. The relatively large frontal bones (*fr.*), which meet in a very wavy median suture, thus extend far backwards; and the squamosals (*sq.*) are likewise much extended, forming the roof of a large post-temporal cavity, of which the opisthotic (Text-fig. 33 B, *opo.*) makes the floor behind. The small postfrontals or sphenotics (*ptf.*) also seem to be exposed on the cranial roof. In the second specimen (B. M. no. P. 9844) there is a pair of large alisphenoids, pierced by several foramina; and the comparatively small but well-ossified orbitosphenoids meet below, completely enclosing the anterior extension of the cerebral cavity.

[1] P. Asmuss, 'Ueber *Aspidorhynchus*,' Archiv für Biontologie, vol. i. (1906), p. 55, pl. vi, and text-fig. 1.

1. Aspidorhynchus fisheri, Egerton. Plate XX, figs. 1—4.

1854. *Aspidorhynchus fisheri*, P. M. G. Egerton, Ann. Mag. Nat. Hist. [2], vol. xiii, p. 434.
1855. *Aspidorhynchus fisheri*, P. M. G. Egerton, Figs. and Descripts. Brit. Organic Remains (Mem. Geol. Surv.), dec. viii, no. 6, pl. vi.
1880. *Aspidorhynchus fisheri*, A. Günther, Introd. Study of Fishes, p. 369, fig. 146.
1895. *Aspidorhynchus fisheri*, A. S. Woodward, Catal. Foss. Fishes, Brit. Mus., pt. iii, p. 425.

Type.—Nearly complete fish; Dorset County Museum, Dorchester.

Specific Characters.—A slender species attaining a length of about 40 cm. Maximum depth of trunk equalling about half the length of the head with opercular apparatus, which is comprised nearly four-and-a-half times in the total length of the fish. Cranium rapidly tapering to the acute rostrum, which is produced in advance of the mandible to an extent equalling one-third of the total length of the cranium; cranial bones and cheek-plates ornamented with fine granulations, which are fused into longitudinal rugæ on the rostrum and sometimes fused or raised on ridges on part of the cranial roof; mandible and opercular bones almost smooth, the former only marked by the openings of the slime-canal; presymphysial bone very short, scarcely longer than deep; mandibular teeth comparatively stout. Pelvic fins arising midway between the pectorals and the caudal. Scales smooth or very feebly rugose, except those of the dorsal region, which are marked with longitudinal rugæ.

Description of Specimens.—The type specimen (Pl. XX, fig. 1) shows the general proportions of the fish, but it is much broken and flaked, and various details of structure are better seen in the other specimens. It exhibits portions of all the fins, indicating their relative positions. It also displays well the arrangement of the scales.

The rostrum, which is closely ornamented with longitudinal ridges, is apparently complete in the type specimen and must have occupied about one-third of the total length of the skull. The rest of the skull, however, and the mandible are better seen in a smaller fossil in the British Museum (enlarged in Pl. XX, fig. 2). The greater part of the cranial roof is formed by the large frontal bones ($fr.$), which are united in a very wavy median suture, and are only slightly excavated at the outer edge by the orbits. They are externally ornamented by fine tubercles, which are arranged in radiating lines, sometimes on ridges, sometimes themselves fused into short ridges. They are also marked along each outer border by the groove and series of pores forming the openings of the slime-canal. The parietals ($pa.$) are short and broad, crossed behind by the groove of the transverse slime-canal, and ornamented with tubercles which are more or less fused into antero-posterior ridges. A crushed mass ($x.$) behind the parietals is probably to be interpreted as the fused epiotics. A slender parasphenoid ($pas.$) is seen crossing the orbit. Of the cheek-plates, there are traces of the large postorbitals, ornamented with a few sparse tubercles; and above them is the characteristic triangular postero-superior circumorbital, similarly ornamented and traversed by the deep groove of the slime-canal with its short posterior branchlets. An antero-superior circumorbital, also apparently of triangular shape, is finely tuberculated; but a small anterior circumorbital, traversed by the slime-canal, is smooth. Remains of an ossified sclerotic ring are distinct.

As shown both by the type specimen and by the original of Pl. XX, fig. 2, the hyomandibular (*hm.*) is an expanded lamina strengthened in the middle by a vertical rod-shaped ridge; and there is a small symplectic (*sy.*) of the *Amia*-type, widest at the upper end, behind the thin, fan-shaped quadrate (*qu.*). The entopterygoid is covered with a cluster of minute, almost granular teeth. The ectopterygoid seems to have borne very small conical teeth. The long and slender maxilla (*m.x.*) is gently curved downwards behind, where it is overlapped by a single small supramaxilla (*smx.*). Its external face is smooth, and its upper border, just in front of the orbit, rises as usual into a broad laminar process. Its oral border bears a row of small conical teeth, which are smallest and curved forwards in the hinder portion. The premaxilla (*pmx.*) seems to have been fused with the base of the rostrum, and its recurved conical teeth diminish in size forwards. The mandible is deep in proportion to its length, and its articulo-angular portion (*ag.*) is very short. The dentary (*d.*) is about three times as deep at its hinder end as at its truncated symphysis. The outer face both of the angular and of the dentary is nearly smooth, only marked by the conspicuous slime-canal, which is defined above by a ridge and fringed below by a series of short branchlets. The oral border of the dentary bears a single regular series of large, smooth, and stout conical teeth, which are flanked near the symphysis by a short row of comparatively small conical teeth. Similar small teeth are also seen in the type specimen on the upper edge of the displaced splenial bone. The presymphysial bone (*ps.*) is imperfect in the type specimen, but well preserved in the original of Pl. XX, fig. 2. It is triangular in shape, not much longer than deep, and nearly smooth on its outer face. Its oral border bears a row of smooth conical teeth, which decrease in size forwards.

The operculum (*op.*), which must have been about as broad as deep, is nearly smooth, only marked by some sparse minute tubercles. The other opercular bones and the branchiostegal rays have not been well seen, but the preoperculum seems to have had the usual triangular expansion of its lower portion. As already noted by Egerton, the removal of the operculum in the type specimen exposes remains of the slender branchial arches, which bear a widely-spaced series of slender gill-rakers (fig. 1*a*).

The smooth cylindrical vertebral centra are crushed flat in the fossils but seem to have been about as long as deep throughout the whole length of the axis.

Traces of short slender ribs occur both in the type specimen and in the original of Pl. XX, fig. 2; and the other vertebral arches are also evidently delicate.

Of the pectoral arch only the clavicle has been clearly observed. This bone (Pl. XX, fig. 2, *cl.*) is much arched, with the upper and lower limbs about equal in length, but the lower the wider. Its exposed portion is very narrow, smooth only at its hinder margin, conspicuously marked by fine longitudinal ridges anteriorly. The hinder border is slightly notched at the insertion of the pectoral fin, where there are traces of a well-ossified scapula or coracoid. There appear to be no specially modified postclavicular scales. The pectoral fin (*pct.*), of which only the base is known, comprises at least 10 smooth rays, of which most are very broad and deeply overlapping. The position of the small pelvic fins is indicated in the type specimen (Pl. XX, fig. 1, *plv.*); but it can only be stated that their rays are smooth and broad, with distant articulations at the distal end. The dorsal (*d.*) and caudal fins are known only by fragments in the same specimen; but the anal fin (*a.*) is better preserved, exhibiting at least 20 rays, of which the length of the foremost is nearly equal to the depth of the tail at its insertion. The anal fin-rays are smooth, divided, and distantly articulated in their distal half, and apparently less crowded than the rays of the paired fins. The foremost ray seems to bear traces of a fringe of minute fulcra.

The total number of transverse series of scales in the type specimen seems to have been slightly more than seventy, and they are seen to be arranged as in the typical species of *Aspidorhynchus* from the Lithographic Stone of Bavaria. Those of the abdominal region are best displayed in outer view in a fragment of which part is represented in Pl. XX, fig. 3. All the scales are deeply overlapping, and their hinder border is not serrated though sometimes slightly wavy. The rhombic dorsal scales throughout the trunk (Pl. XX, figs. 1*b*, 4) are covered with conspicuous smooth ridges, variously wavy and sometimes bifurcating, which run diagonally. The other scales, though also well enamelled, are only faintly and finely rugose. In each transverse series of the abdominal region there are two of the ornamented rhombic scales at the upper end, and there seems to have been a similarly ornamented median dorsal ridge-scale. The two principal scales of the flank are about equal in depth, and the upper of these is crossed near its forwardly curved upper end by the slime-canal, which notches the hinder border and occasionally opens by a pore on the middle of the scale. The next lower scale is almost square, less than half as deep as the principal flank-scales. Six, sometimes seven, very narrow scales complete the series below, the lowest being a sharp-edged ridge-scale without any serration. In the caudal region the flank-scales gradually become less deepened, until on the caudal pedicle they are rhombic and nearly uniform in size. There appear to be no enlarged ridge-scales. When exposed from within (Pl. XX, figs. 1, 2), all the scales of the abdominal region are shown to be united by a very large peg-and-socket articulation, and their inner face is strengthened by a wide vertical ridge. In the caudal region this articulation gradually disappears.

Horizon and Locality.—Middle Purbeck Beds: Swanage, Dorset.

Genus BELONOSTOMUS, Agassiz.

Belonostomus, L. Agassiz, Neues Jahrb. für Min., etc., 1834, p. 388.
Ophirhachis, O. G. Costa, Ittiol. Foss. Ital.; 1856, p. 13.
Diphyodus, L. M. Lambe, Contrib. Canadian Palæont., vol. iii, pt. ii, 1902, p. 30.

Generic Characters.—As *Aspidorhynchus*, but mandible almost or quite as long as the rostrum, and all the scales of the lateral line deeper than those immediately beneath.

Type Species.—*Belonostomus sphyrænoides* (L. Agassiz, Poiss. Foss., vol. ii, pt. ii, 1844, pp. 140, 297, pl. xlvii *a*, fig. 5), from the Lithographic Stone (Lower Kimmeridgian) of Bavaria.

1. Belonostomus hooleyi, sp. nov. Plate XXI, figs. 1—3.

Type.—Imperfect principal scale of flank; collection of Reginald W. Hooley, Esq.

Specific Characters.—Principal deepened scale of flank smooth, sometimes with traces of large elongated tubercles near the anterior border, and marked by one conspicuous vertical groove; the posterior border very coarsely and irregularly crenulated and pectinated. Dorsal scales ornamented with a few large, low, and irregular short longitudinal ridges. Roof of skull with similar coarse ornament.

Description of Specimens.—The type scale (Pl. XXI, fig. 1) is incomplete below, and most of its smooth outer surface has been flaked away; but it is well preserved at the forwardly bent upper end, where the slightly tumid anterior half is clearly separated by a sharp vertical line from the flatter posterior half. At the upper end the posterior border seems to be entire, but below this it is very coarsely and irregularly crenulated, each prominence being bluntly pointed and forming a slight horizontal ridge.

The same form of flank-scale is distinguishable in the fragmentary pyritised remains of a fish in the Mantell Collection, which also includes the hinder half of a cranium and traces of vertebræ. The flattened roof of the skull must have been very coarsely ornamented with irregular rounded ridges and low elongated tubercles. All the otic bones and apparently the alisphenoid are well ossified; and an anterior vertebral centrum, evidently fused with the basioccipital, is more extensively ossified than the ordinary vertebral centra of *Aspidorhynchus*. Another centrum among the scales shows that it is pierced in the middle for a persistent strand of the notochord. The rhombic dorsal scales (Pl. XXI, fig. 2) are as coarsely ornamented with rounded ridges as the bones of the cranial roof. They are deeply overlapping, and their irregular ornamental ridges tend to be diagonal in direction.

Remarks.—The fragmentary remains thus described are referred to *Belonostomus* rather than to *Aspidorhynchus*, on account of the stout ossification of the vertebral centra and the vertical grooving of the principal flank-scales. They are distinguished from the remains of all known species by the peculiar crenulation of the posterior border of the principal flank-scales. The specific name now proposed for them is given in honour of Mr. Reginald W. Hooley, F.G.S., whose important discoveries in the Wealden of the Isle of Wight are well known.

Horizon and Localities.—Wealden: Isle of Wight, probably also Sevenoaks, Kent (small scale in Sedgwick Museum, Cambridge, shown enlarged in Pl. XXI, fig. 3).

Family PHOLIDOPHORIDÆ.

Genus **PHOLIDOPHORUS**, Agassiz.

Pholidophorus, L. Agassiz, Neues Jahrb. für Min., etc., 1832, p. 145.
(?) *Microps*, L. Agassiz, Poiss. Foss., vol. ii, pt. i, 1833, p. 10.

Generic Characters.—Trunk fusiform, not much deepened, and head relatively large. External bones smooth or delicately ornamented with rugæ and tuberculations; slime-canal on cheek-plates branched; maxilla more or less arched, the oral margin convex or nearly straight, with small conical or styliform teeth; mandibular teeth slightly larger, in single series. Lower expansion of preoperculum marked by radiating furrows; subopercolum large, but smaller than the trapezoidal operculum, from which it is divided by an oblique suture; interoperculum triangular and well forwards; branchiostegal rays numerous. Pleurocentra and hypocentra round notochord fused or separate. Fulcra present on all the fins. Pectoral fins of moderate size, but larger than the pelvic pair; dorsal and anal fins triangular, not extended, the former opposite to or arising somewhat behind the pelvic fins; caudal fin deeply forked. Squamation complete; scales thin and deeply overlapping, usually with an inner rib and peg-and-socket articulation, and the external ganoine smooth or feebly ornamented; principal flank-scales deeper than broad, ventral scales in part broader than deep; no enlarged series of ridge-scales. Lateral line conspicuous.

Type Species.—*Pholidophorus bechei* (L. Agassiz, Poiss. Foss., vol. ii, pt. i, 1844, p. 272, pl. xxxix, figs. 1—4), from the Lower Lias of Lyme Regis, Dorset. Additional details in Catal. Foss. Fishes, Brit. Mus., pt. iii (1895), p. 450, pl. xii, figs. 1, 2.

Remarks.—Representatives both of the pectinate-scaled and of the smooth-scaled forms of *Pholidophorus* occur in the Purbeck Beds. They are remarkable for the stoutness of the fulcra on the fins.

1. **Pholidophorus ornatus**, Agassiz. Plate XX, figs. 5—8; Plate XXI, fig. 4; Text-figure 34.

1843-44. *Pholidophorus ornatus*, L. Agassiz, Poiss. Foss., vol. ii, pt. i, p. 280, pl. xxxvii, figs. 6, 7.
1855. *Pholidophorus ornatus*, P. M. G. Egerton, Figs. and Descripts. Brit. Organic Remains (Mem. Geol. Surv.), dec. viii, no. 4, p. 1, pl. iv, fig. 3.
1895. *Pholidophorus ornatus*, A. S. Woodward, Catal. Foss. Fishes, Brit. Mus., pt. iii, p. 471.

Type.—Imperfect tail; British Museum.

Specific Characters.—Attaining a length of about 20 cm. Length of head with opercular apparatus nearly equal to the maximum depth of the trunk and occupying between one-quarter and one-fifth of the total length of the fish. Head and opercular bones very feebly rugose or smooth; maxilla not much arched; teeth obtusely pointed. Fin-rays smooth and stout, and fulcra comparatively large. Pelvic fins arising midway between the pectoral and anal fins, and opposed to the dorsal fin, of which the length of the longest ray about equals the depth of the trunk at its insertion. Scales ornamented with coarse oblique pectinations ending in sharp serrations; about six flank-scales in each abdominal transverse series deeper than broad; lateral line marked by notches on the scales in the abdominal region, forming a sharp smooth ridge on the scales in the hinder part of the caudal region.

Description of Specimens.—The type specimen in the Mantell Collection (Pl. XX, fig. 5) exhibits only the imperfect caudal region, and is very unsatisfactorily described and figured by Agassiz, *loc. cit.* The thin cylindrical vertebral centra (*v.*) are seen, though crushed, near the base of the caudal fin, while the comparatively short and stout neural and hæmal arches are shown to be sharply inclined backwards and thus deeply overlapping. A fragment of the pelvic fin (*plv.*) proves that it was rather large. The small anal fin (*a.*) bears a close series of short and slender fulcra. The dorsal fin is absent, though false appearances of it were misinterpreted and described by Agassiz. The stout rays of the forked caudal fin are enamelled, and with rather distant articulations. Its fringing fulcra are small and in very close series. The scales exhibit the characteristic pectination of the posterior border even to the base of the caudal fin; and there is one enlarged, elongated dorsal ridge-scale, with faint rugose ornament, at the origin of the upper caudal lobe. Traces of a smooth longitudinal keel are seen on some of the scales of the lateral line.

Other specimens now exhibit nearly all the principal characters of the genus and species, and one small fish (Pl. XX, fig. 6), discovered by Mr. Alexander J. Hogg, is especially fine, only crushed and incomplete towards the end of the tail.

The abdominal region must have been rather stout, not much laterally compressed, for it is often exposed from below (Pl. XXI, fig. 4) or distorted by crushing in the fossils. Its ventral face must have been nearly flat.

The postorbital part of the cranial roof is somewhat wider than long, and the frontal region between the orbits is also comparatively wide. Except the narrow overlapped margin at its straight occipital border, the roof is closely ornamented with a very fine and inconspicuous rugosity; the short parietals and squamosals are crossed by a groove for the transverse slime-canal, and the frontals are pitted along the outer margin by the openings of the longitudinal slime-canal (B. M. no. 28446). A large elongate-oval nasal bone, also ornamented, occurs in front of the frontal (B. M. no. P. 10011). There is much ossification in the otic region; and the parasphenoid bone is stout, bearing an elongated cluster of minute teeth between and just in front of its lateral processes (B. M. no. 28445).

The large orbit is surrounded by a ring of thin, smooth cheek-plates, traversed

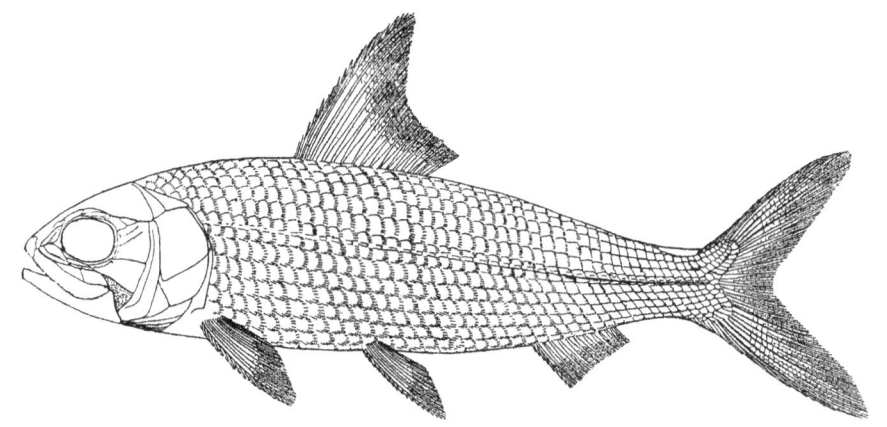

Fig. 34.—*Pholidophorus ornatus*, Agassiz; restoration, about half nat. size.—Middle Purbeck Beds; Swanage. The cheek-plates partly restored from *Pholidophorus micronyx*, Ag.

by a very large slime-canal, from which a few long branchlets radiate on the postorbitals and preorbitals. There is an ossified sclerotic ring.

In the mandibular suspensorium the hyomandibular is much expanded, and the small quadrate is cleft behind as if for clasping a symplectic. The entopterygoid is a thin lamina of bone, and a cluster of minute teeth in B. M. no. 28446 may have belonged to it. The maxilla (Pl. XX, fig. 6 a, *mx.*) and premaxilla (*pmx.*) are comparatively stout bones with a smooth or only faintly rugose outer face. The maxilla is a narrow band of bone curved upwards at its hinder end, where it is deepest, and not much contracted at its anterior end, where it bears a slender upwardly-directed process for attachment to the inner face of the premaxilla and doubtless also to the palatine (especially well seen in Mus. Pract. Geol. no. 28438). Its oral border is nearly straight, bearing a single close regular series of small and blunt styliform teeth; its upper border, throughout the greater part of its length, is overlapped by two thin and nearly smooth supramaxillary plates (*smx.*), the posterior being the smaller and triangular in shape with a short antero-superior process, the anterior elongate-triangular slowly tapering to a point in front. The premaxilla has an oral border scarcely more than one-quarter as long as that of the maxilla, but its styliform teeth are somewhat larger; its upper portion near the middle line of the snout rises into a process for articulation with the ethmoidal region of the skull. The mandible tapers gradually from the low coronoid elevation behind to the bluntly pointed symphysis in front. The articulo-angular bone (*ag.*) is very short, with its finely rugose ornament slightly more conspicuous than that of the dentary (*d.*), which is marked chiefly by the row of large pores along the course of the slime-canal. Its blunt teeth, which are somewhat stouter than those of the maxilla, seem to be in more than one series at the symphysis.

The opercular bones are so thin that they are usually much broken in the fossils, and their fine rugose ornament is so delicate that it is often destroyed. The operculum (Pl. XX, fig. 6; Pl. XXI, fig. 4, *op.*) is a little deeper than wide and slightly more than twice as deep as the subopereulum (*sop.*), which bears a conspicuous sharply pointed antero-superior process. The triangular interoperculum (*iop.*) is relatively large. The preoperculum (*pop.*), which tapers upwards nearly to the upper limit of the operculum, is a narrow plate, not much expanded at its blunt angle, and almost without a horizontal limb. It is traversed by a very large slime-canal, from which a few branchlets radiate backwards over the lower expansion. The upper branchiostegal rays (*br.*) are large and finely rugose, and all these rays seen in the fossils, not less than 16 pairs in regular series, are rather broad laminæ.

The delicate vertebral rings in the caudal region, with their short recumbent neural and hæmal arches, in the type specimen, have already been mentioned. Other specimens show that the centra are represented only by rings throughout the vertebral axis; and in some cases a ring appears to consist of two separate parts, a pleurocentrum above and a hypocentrum below (B. M. no. P. 3605 *a*). The ribs are short and slender (B. M. no. 43038).

The supratemporal and post-temporal plates have not been well observed, but they are evidently thin and very faintly rugose. The supraclavicle (Pl. XXI, fig. 4, *scl.*) is about three times as deep as wide, and crossed obliquely by the large slime-canal, above which its posterior margin is enamelled and pectinated. The sigmoidally bent clavicle (*cl.*) is nearly smooth in its narrow exposed portion, but rises into a triangular rugose boss just beneath the notch for the insertion of the pectoral fin, and then passes below into a large smooth expansion where it is covered. The postclavicular plates, or perhaps anterior scales, are smooth for the greater part of their width, but pectinated and serrated at their posterior margin. The pectoral fin (Pl. XXI, fig. 4, *pct.*) comprises about 20 rays, all articulated and divided distally, and the foremost fringed with a close series of small, deeply overlapping fulcra. The pelvic fins (*plv.*), which are inserted widely apart, are not much shorter than the pectorals; but each comprises only about 10 rays, the foremost fringed with fulcra which are larger than those of the pectoral and especially stout at the base. In the original of Pl. XX, fig. 6, the foremost pelvic fin-ray is ornamented with longitudinal flutings of enamel. The dorsal fin is best preserved in the same specimen, with about 16 smooth rays which rapidly decrease in length backwards. In a larger specimen (Mus. Pract. Geol. no. 28438) there are three or four slender basal fulcra besides conspicuous fringing fulcra above, and some of the fin-rays are marked by a longitudinal line of enamel. The anal fin, which is well behind the dorsal, is comparatively small with 9 or 10 rays, best seen, though crushed, in the type specimen (Pl. XX, fig. 5, *a*.). Its fulcra are very slender and deeply overlapping, and its broad articulated rays bear some smooth longitudinal strips of enamel. The caudal fin is also best seen in the type specimen, as already described (p. 102). All the fin-rays are much expanded laminæ, obliquely overlapping: their appearance of stoutness, therefore, in the fossils depends upon the direction in which they are exposed.

Nearly all the scales are conspicuously ornamented with coarse plications ending in sharp prominent serrations. As counted along the course of the lateral line they are regularly arranged in from 40 to 45 transverse series; and the series above the origin of the pelvic fin comprises 13 or 14 scales, of which the seventh or eighth is crossed by the lateral line. On the flank of the abdominal region about six scales in each series are deeper than broad, that crossed by the lateral line and the one immediately below being especially deepened. The two or three lowest ventro-lateral scales are not so deep as broad; and one of these seems to be bent along its long axis to form the edge of the ventral surface of the fish. All these scales are very conspicuously ornamented by the oblique pectinations throughout their depth; but the postero-superior angle is sometimes rounded or truncated, while on the less deepened scales of the caudal region the pectinations gradually become restricted to the lower part and sometimes on the caudal pedicle disappear. Among the rhombic scales of the ventral surface (Pl. XX, fig. 8) and the dorsal border, there is also sometimes a tendency to similar réduction of the ornament. On the scales of the abdominal region the lateral line is marked only by a posterior notch and an occasional perforation; but on the rhombic scales in the hinder half of the caudal region (Pl. XX, fig. 7), it forms a sharp smooth ridge, which is also occasionally perforated in the middle. There is slight irregularity in the arrangement of the scales at the insertion of the paired fins; and on the inner side of each pelvic fin (Pl. XXI, fig. 4, *plv.*) there is a peculiar row of two small scales and a posterior large elongate-oval scale, less ornamented than usual. As shown by the type specimen (Pl. XX, fig. 5) a large elongate-ovoid ridge-scale, finely rugose but not serrated, occurs on the upper border of the caudal pedicle at the origin of the caudal fin. There are no prominent ventral ridge-scales (Pl. XXI, fig. 4). All the principal scales are united by a peg-and-socket joint, which is strengthened by a low vertical ridge on their inner face (Pl. XX, fig. 7a).

Horizon and Localities.—Middle Purbeck Beds: Swanage; Upway, near Weymouth.

2. **Pholidophorus granulatus,** Egerton. Plate XXI, figs. 5, 6.

1854-55. *Pholidophorus granulatus,* P. M. G. Egerton, Ann. Mag. Nat. Hist. [2], vol. iii, p. 434; and Figs. and Descripts. Brit. Organic Remains (Mem. Geol. Surv.), dec. viii, no. 4, pl. iv, figs. 1, 2.
1895. *Pholidophorus granulatus,* A. S. Woodward, Catal. Foss. Fishes, Brit. Mus., pt. iii, p. 470.

Type.—Imperfect fish; Dorset County Museum, Dorchester.

Specific Characters.—A very robust species, attaining a length of about 30 cm.; maximum depth of trunk equalling about one-third of the total length of the fish. Head and opercular bones finely tuberculated or rugose, the tuberculations extending over the dorsal scales of the abdominal region. Fin-rays smooth and stout, and fulcra comparatively large. Pelvic fins arising far in advance of the middle point of the trunk, and the dorsal fin opposed to them. Scales large, ornamented with very fine oblique ridges, which radiate slightly and end at the hinder margin in delicate serrations; several series of flank-scales deeper than broad; lateral line inconspicuous.

Description of Specimens.—The type specimen displays well the squamation of the trunk and the paired fins; but it is deepened a little by crushing in the ventral region, and the head bones are broken and displaced. Of the median fins it exhibits only fragments. The other known specimens are still more imperfect, but they show several of the most important characters of the genus and species. One of these specimens (Pl. XXI, fig. 5) is interesting as proving that in the abdominal region there is a narrow flattened ventral face between the paired fins.

The roof of the skull, partly shown in Pl. XXI, fig. 6, is broad and not much arched from side to side. The parietals (*pa.*) and squamosals (*sq.*) are not more than one-third as long as the frontals, and their external face is closely ornamented with radiating rows of fine tubercles. They are crossed by a groove for the transverse slime-canal, and their hinder margin is deeply overlapped by the supratemporals. Each frontal (*fr.*) is nearly twice as wide behind as in front, slightly excavated for the large orbit at the outer border, and bifurcating at the front border, where the antero-external angle is less produced than the antero-internal angle, which must have rested on the mesethmoid. The median frontal suture is deeply jagged between the middle of the orbits, but otherwise only slightly wavy. A narrow supraorbital margin is nearly smooth, but the greater part of the frontal is closely ornamented by radiating rows of fine tubercles, which are fused into coarser ridges on the anterior bifurcating portion.

The longitudinal slime-canal is marked by a few pores. Remains of the supratemporal and post-temporal plates behind are also closely and finely tuberculated.

The opercular apparatus is shown in inner view in Pl. XXI, fig. 5. The operculum (*op.*) is narrow above owing to the rounding of the postero-superior angle, and its depth to the middle of the lower border equals its maximum width. The large suboperculum (*sop.*) is half as deep as the operculum, and bears the usual antero-superior prominence. The preoperculum is much expanded at its angle, and the interoperculum is relatively small. The opercular bones and remains of branchiostegal rays in the type specimen are finely granulated.

Although the neural and hæmal arches in the vertebral axis are well ossified, the centra must have been extremely delicate, for there is scarcely a trace of them in the known specimens.

The scales are regularly arranged in about 45 transverse series; and there are 16 scales in the series above the origin of the pelvic fins, the ninth from below being traversed by the slime-canal of the lateral line. On the flattened ventral surface of the abdominal region they are rhombic and nearly equilateral, deeply overlapping, and ornamented in their exposed portion with fine radiating ridges, which end behind in serrations and pass forwards into tuberculations. On the lateral edge of this ventral area each scale is bent along its antero-posterior diagonal. On the flank of the abdominal region the dorsal and ventral scales are also rhombic and nearly equilateral; but in each series eight or nine scales are deeper than broad, the deepest being those of the lateral line and the three horizontal rows beneath it. In the anterior series (Pl. XXI, fig. 5a) these scales are very deeply overlapping, and all are conspicuously ornamented with the fine radiating ridges, which end at the hinder border in serrations. On the principal scales the ridges cover nearly the whole of the exposed surface, leaving only a small triangular rugose or tuberculated area near the lower border; but on most of the abdominal scales they pass forwards into tuberculations or rugæ. Further back and on the caudal region (fig. 5b) the fine radiating ridges gradually become shorter, leaving the rest of the scale smooth; until towards the end of the caudal pedicle both ridges and serrations disappear. In the anterior part of the abdominal region (Pl. XXI, fig. 5a) the lateral line is marked only by a slight ridge and posterior notch on the scale; but further back and in the caudal region (fig. 5b) it opens by a prominent perforation on most of the scales. In all the scales of the abdominal region the peg-and-socket articulation is large, while the inner rib is broad and prominent (Pl. XXI, fig. 6a); in the hinder half of the caudal region (fig. 6b) both the articulation and the inner rib disappear. There are enlarged scales round the anus near the origin of the anal fin.

The position of the fins is shown in the type specimen, but they are imperfect in all known specimens. The pectoral fins, which comprise about 18 rays, are not much more elevated than the pelvic fins, which comprise 10 or 12 rays and are fringed with stout fulcra. The pelvic pair is inserted nearer to the pectorals than to the anal, directly opposite the origin of the dorsal fin, which begins with 5 or 6 very stout but small basal fulcra, more or less enamelled, and comprises not less than 12 rays, perhaps 2 or 3 more. The anal fin must have been comparatively small; in the original of Pl. XXI, fig. 5, there are the bases of 9 rays, with traces of stout basal fulcra.

Horizon and Locality.—Middle Purbeck Beds: Swanage, Dorset.

3. Pholidophorus purbeckensis, Davies. Plate XXII, figs. 1—3.

1887. *Pholidophorus purbeckensis*, W. Davies, Geol. Mag. [3], vol. iv, p. 337, pl. x, figs. 2—4.
1888. *Pholidophorus purbeckensis*, W. Davies, in R. Damon, Geol. Weymouth, ed. 3, Suppl. pl. xix, fig. 1.
1895. *Pholidophorus purbeckensis*, A. S. Woodward, Catal. Foss. Fishes, Brit. Mus., pt. iii, p. 460.

Type.—Imperfect fish; British Museum.

Specific Characters.—Attaining a length of 8—10 cm. Length of head with opercular apparatus about equal to the maximum depth of the trunk and nearly one-quarter of the total length of the fish. Head and opercular bones smooth or feebly rugose; teeth stout and sometimes slightly recurved. Fulcra smooth, unusually large and stout on all the fins. Pelvic fins inserted about midway between the pectoral and anal fins; dorsal fin opposed to the space between the pelvic and anal fins, with about 10 rays, the length of the foremost and longest somewhat less than the depth of the trunk at its insertion; anal fin also with about 10 rays, as elevated as the dorsal; length from insertion of anal fin to that of caudal fin about equal to maximum depth of trunk. Scales large and smooth, the hinder margin not serrated; those of the lateral line much deepened, some four times as deep as broad; two scales above and below those of the lateral line also moderately deepened; lateral line forming only a feeble ridge.

Description of Specimens.—The type specimen (Pl. XXII, fig. 1), obtained by Mr. Robert Damon from the Lower Purbeck of the Isle of Portland, lacks the greater part of the head, but displays well the contour of the trunk in side-view, with fragments of all the fins. It also shows satisfactorily the squamation. Another specimen, much splintered by crushing, also figured by Davies (*loc. cit.* 1887, pl. x, fig. 2), gives the proportions of the head and caudal fin, and displays the characteristic large fulcra on the dorsal fin. A still finer specimen, with all the fins, is represented somewhat magnified in Pl. XXII, fig. 2, with a further enlargement of the fin-fulcra in fig. 2a. A dwarf example evidently of the same species from a corresponding horizon at Teffont, Wiltshire, is shown enlarged twice in Pl. XXII, fig. 3.

As shown in a specimen figured by Davies (1887, pl. x, fig. 2), and in another from Swanage (B.M. no. P. 12347), the postorbital part of the cranial roof is very feebly but coarsely rugose. So far as can be observed in other specimens, the cheek-plates, mandible, and opercular bones are for the most part smooth, but bear occasional prominent tubercles. The cheek-plates seem to be arranged as ordinarily in *Pholidophorus*, but they are often much broken by crushing (as in the original of Pl. XXII, fig. 3). The large orbit as usual is crossed by a very stout parasphenoid bone; and an inner bone of the mouth, perhaps the vomer, bears a few stout conical teeth. The mandibular suspensorium is inclined forwards below, so that the quadrate articulation is beneath the hinder half of the orbit. The maxilla is gently arched, with comparatively small teeth. About half the length of the mandibular ramus is occupied by its elevated coronoid portion.

As shown in the type specimen (Pl. XXII, fig. 1, *pop.*) the narrow preoperculum is gently curved forwards below and not much expanded at the angle, where it is grooved and pitted by the slime-canal but otherwise smooth. As shown from within in Pl. XXII, fig. 3, the suboperculum (*sop.*) is scarcely half as deep as the operculum (*op.*), while the interoperculum (*iop.*) is large, long, and narrow. The branchiostegal rays are short and broad; and there seems to be a long and narrow gular plate in the specimen figured by Davies (1887, pl. x, fig. 2), though this interpretation is not certain.

The axial skeleton of the trunk is always more or less obscured by the squamation, but a close series of vertebral centra in the form of delicate ossified cylinders is seen in side-view in B.M. no. P. 8379, and an isolated centrum occurs in end-view in the same specimen. Traces of similar centra and short delicate ribs are also seen in the original of Pl. XXII, fig. 2.

In the pectoral arch the supraclavicle (Pl. XXII, fig. 1, *scl.*) is about three times as deep as broad, with the outer face and hinder margin smooth. The narrow exposed portion of the clavicle is marked by longitudinal ridges or plications. Two postclavicular scales are clearly seen above the pectoral fin both in the original of Pl. XXII, fig. 3, and in B.M. no. 40635, the lower being about as broad as deep, the upper much deeper than broad, and tapering upwards. In the pectoral fin (Pl. XXII, fig. 2) the foremost ray is much stouter than the others, and seems to have borne large fulcra. The pelvic fins are not much smaller than the pectorals, and their foremost ray bears at least six large and deeply overlapping smooth fulcra, which extend nearly to its distal end. The dorsal and anal fins, the former just in advance of the latter, are similar in size and shape, each with about ten divided rays, which rapidly decrease in length backwards, and of which the foremost is fringed nearly to the end with eight or nine large and deeply overlapping smooth fulcra. The relatively large forked caudal fin, with nearly twenty rays, is similarly fulcrated, but the dozen fulcra above and below rapidly decrease in size distally and do not reach the extremity of the foremost ray. In all the fin-rays the segments between the articulations are somewhat longer than broad.

All the scales are smooth, with no posterior serrations. They are strengthened on their inner face with a stout vertical ridge, and those of the abdominal region are united by a large peg-and-socket articulation. They are arranged in nearly 40 regular transverse series; and above the origin of the pelvic fin the series comprises 11 or 12 scales, of which the lateral line traverses the seventh or eighth from the ventral border. In the abdominal region (Pl. XXII, fig. 1a), the scales of the lateral line are from four to three times as deep as broad; the next scales above and below these are about twice as deep as broad; the next above and below are still somewhat deeper than broad; those dorsally and ventrally are smaller and nearly equilateral, or even broader than deep. The very narrow flattened ventral face of the abdominal region is covered with small rhombic scales; and there seem to have been slightly enlarged and modified scales, with two or three posterior denticulations, round the anus. In the caudal region (Pl. XXII, fig. 1b) the scales of the lateral line still remain somewhat deeper than broad, and they are crossed very obliquely by the slime-canal. The other caudal scales are rhombic and nearly equilateral, with a tendency to the rounding of the hinder and upper margins. There is no conspicuously enlarged ridge-scale on the caudal pedicle above or below, but there seem to be two or three short and broad dorsal ridge-scales. The lateral line is marked only by a faint smooth ridge.

Horizons and Localities.—Lower Purbeck Beds: Portland, Dorset; Teffont, Wiltshire. Middle Purbeck Beds: Swanage, Dorset.

4. Pholidophorus brevis, Davies. Plate XXII, figs. 4, 5.

1887. *Pholidophorus brevis*, W. Davies, Geol. Mag. [3], vol. iv, p. 338, pl. x, fig. 1.

Type.—Imperfect fish; British Museum.

Specific Characters.—A stout mutation of *P. purbeckensis*, with a comparatively short and robust caudal region.

Description of Specimens.—The type specimen (Pl. XXII, fig. 4), from the Upper Purbeck Beds, is distorted, so that the head is accidentally shortened and the abdominal region deepened, but it evidently represents a shorter and stouter fish than the Lower Purbeckian *P. purbeckensis*. Part of a second specimen, in counterpart in the Egerton and Enniskillen Collections, indicates an equally stout form (Pl. XXII, fig. 5).

In the latter fossil the head appears to exhibit its true shape; and its length to the back of the opercular apparatus clearly equals the maximum depth of the trunk. Traces of a coarse rugosity are seen on the cranial roof. Stout conical teeth, some with a slightly curved apex, occur in both jaws of the type specimen. The suboperculum is less than half as deep as the operculum in both specimens. Crushed obscure remains of the very delicate ring-vertebræ are preserved especially in the counterpart of the specimen figured in Pl. XXII, fig. 5.

The pelvic, dorsal, and anal fins are rather well shown in the type specimen (Pl. XXII, fig. 4), which also retains fragments of the pectorals and caudal. The former appear to agree in all respects with the corresponding fins of *P. purbeckensis*, being only slightly smaller in proportion to the depth of the trunk. The scales are also evidently similar to those of the earlier species just mentioned. In both specimens they are much fractured, and nearly all are exposed from the inner face, displaying the strong vertical rib, with the peg-and-socket articulation in the abdominal region. They are, however, clearly smooth and without posterior serration.

Horizon and Locality.—Upper Purbeck Beds: Upway, near Weymouth, Dorset.

Genus **CERAMURUS**, Egerton.

Ceramurus, P. M. G. Egerton, in Brodie's Fossil Insects, 1845, p. 17.

Generic Characters.—Trunk slender fusiform, and head relatively large; notochord persistent, surrounded with spaced ring-vertebræ; ribs short and delicate. Fin-fulcra few, long and slender. Pectoral and pelvic fins long, but few-rayed; dorsal and anal fins not extended, the former in advance of the latter; caudal fin long, but probably somewhat forked. Squamation absent on flanks, except, perhaps, a rudiment anteriorly; a short series of robust ganoid ridge-scales on both borders of the hinder half of the caudal region.

Type Species.—*Ceramurus macrocephalus*, from the English Purbeck Beds.

1. **Ceramurus macrocephalus**, Egerton. Plate XXII, fig. 7.

1845. *Ceramurus macrocephalus*, P. M. G. Egerton, in Brodie's Fossil Insects, p. 17, pl. i, fig. 2.
1895. *Ceramurus macrocephalus*, A. S. Woodward, Geol. Mag. [4], vol. ii, p. 401; and Catal. Foss. Fishes, Brit. Mus., pt. iii, p. 489.

Type.—Nearly complete fish; British Museum.

Specific Characters.—A slender species, at least 4·5 cm. in length. Length of head with opercular apparatus nearly twice as great as the maximum depth of the trunk, and contained about four-and-a-half times in the total length of the fish. Pectoral and pelvic fin-rays about equal in length; dorsal fin with about 10 rays, opposed to the pelvic pair; anal fin with 8 or 9 rays, completely behind the dorsal. Caudal ridge-scales smooth, those of the upper border especially acuminate, each being produced into a long point.

Description of Specimens.—This species is known only by three specimens: the type discovered in the Middle Purbeck of Dinton, Wiltshire, by the Rev. P. B. Brodie (shown enlarged in Pl. XXII, figs. 7, 7*a*); a more imperfect fish discovered in the Lower Purbeck of Teffont, Wiltshire, by the Rev. W. R. Andrews (B.M. no. P. 9850); and a nearly complete specimen from an unrecorded locality near Weymouth (B.M. no. P. 7506). The type specimen is preserved in counterpart, but the head and caudal fin are crushed and imperfect, while a vein of calcite crosses obliquely the hinder end of the abdominal region. Remains of the cranial roof show that the bones are smooth or only faintly rugose. The frontals (*fr.*) form a symmetrical pair separated by a straight median suture, and excavated laterally by a large orbit. The parietals (*pa.*), which are also a symmetrical pair divided by a straight median suture, are each longer than broad, and marked behind by a large groove for the transverse slime-canal. They are well seen again in inner view in B.M. no. P. 9850. The parasphenoid, which is displaced, is very stout. The mandibular suspensorium must have been inclined forwards, and the small fan-shaped quadrate (*qu.*), with its posterior cleft for the symplectic, is seen in the fossil below the front of the orbit. Remains of the opercular apparatus show that it is smooth.

The vertebral centra are delicate rings, which appear to have been arranged in a spaced series, so that in their crushed condition in the type specimen most of them are exposed in end-view. Some of those in the abdominal region of B.M. no. P. 7506 are in side-view, and shown to be slightly constricted. Towards the end of the caudal region the centra may have been incomplete, and in the upturned extremity they are apparently absent. In advance of the dorsal fin the neural arches are much shorter than their appended spines, which are loosely apposed and seem to have nearly reached the dorsal border. Beneath the dorsal fin the short neural arches bear extremely short spines. The ribs shown in the type, but best seen in B.M. no. P. 9850, are short and delicate throughout the abdominal region. The short neural and hæmal spines in the caudal region are fused with their respective arches, and these also probably with the vertebral rings. Where the axis is upturned within the caudal fin, the hæmal spines are enlarged and thickened as usual. There are no traces of intermuscular bones.

The fins consist of long, slender rays, with distant articulations, and apparently not forked more than once at the distal end. Each is fringed by a few very long and slender fulcra, which are conspicuous by their enamelled surface. The clavicle (*cl.*) is relatively large and much curved, widest below, and with a thin laminar expansion in its anterior concavity. Above the left clavicle, which is displaced backwards in the type specimen, there is a deep and narrow supraclavicle (*scl.*). There also seem to be remains of large, smooth postclavicular scales. Both the pectoral and pelvic fins are crushed close to the body, and display their characteristic fulcra. One of the pelvic fin-supports is seen, slightly expanded proximally, much constricted mesially, more widely expanded distally. The length of the longest rays of the short dorsal fin is greater than the depth of the trunk at their insertion, and the elongated fulcra are especially stout at the base. Nine supports, of which those anteriorly are stouter than the rest and somewhat winged, can be counted in front of the vein of calcite in the type fossil. The anal fin, which is slightly less elevated than the dorsal, comprises 8 or 9 rays, which, when adpressed, nearly reach the base of the caudal fin. Five fulcra are conspicuous on the anterior margin. The caudal fin, known only in the type specimen, is imperfect, and distorted upwards so that its precise shape is uncertain, but it seems to have been forked.

The only traces of scales on the trunk are immediately behind the pectoral arch, but these may represent merely postclaviculars. Robust, smooth, overlapping ganoid ridge-scales, however, are conspicuous on both borders of the caudal pedicle. The dorsal series begins above the middle of the anal fin, the scales all very sharply pointed (Pl. XXII, fig. 7b) and gradually increasing in size until they pass into the upper caudal fulcra. The ventral series comprises only three smaller and less acuminate scales, occupying the hinder half of the space between the anal and caudal fins.

Horizon and Locality.—Middle and Lower Purbeck Beds: Vale of Wardour, Wiltshire. Purbeck Beds: near Weymouth, Dorset.

Genus **PLEUROPHOLIS**, Egerton.

Pleuropholis, P. M. G. Egerton, Figs. and Descripts. Brit. Organic Remains (Mem. Geol. Surv.), dec. ix, 1858, no. 7.

Generic Characters.—Trunk elongate-fusiform, with rounded back and sharp ventral border; upper caudal lobe conspicuous. External bones smooth, or delicately ornamented with rugæ and tuberculations; sensory canal on cheek-plates with branches; mouth very small, with minute teeth; maxilla more or less arched, the oral margin convex. Vertebral centra annular; ribs short and delicate. Fulcra present on all the fins. Pelvic fins well developed, but smaller than the pectorals; dorsal and anal fins longer than deep, opposite; caudal fin forked. Squamation complete; scales thick and moderately overlapping; those of the middle of the flank excessively deepened, covering nearly the whole of it, each strengthened within by a broad rib and exhibiting a peg-and-socket articulation; dorsal and ventral scales few, relatively small and rhomboidal. Lateral line deflected, passing down the second or third deepened flank-scale and then traversing the uppermost row of small ventral scales.

Type Species.—*Pleuropholis attenuata*, from the English Middle Purbeck Beds.

Remarks.—The osteology of the head in *Pleuropholis* is still very imperfectly known, but its cheek-plates seem to resemble those of *Pholidophorus* in being only in a single series. The most remarkable feature of the genus is the deflection of the lateral line; but it must be added that in one species from the Lithographic Stone of Bavaria Miss Mary S. Johnston has also discovered slight traces of a slime-canal on the deepened flank-scales of the caudal region (Geol. Mag. [5], vol. vi, 1909, p. 311), and there are similar traces in *Pleuropholis serrata* described below (p. 121).

Pleuropholis occurs chiefly in the Lithographic Stone (Lower Kimmeridgian) of Bavaria and France, but is also known by fragments in the Wealden of Belgium (R. H. Traquair, " Les Poissons Wealdiens de Bernissart," Mém. Mus. Roy. Hist. Nat. Belg., vol. v, 1911, p. 45, pl. ix, figs. 1—3).

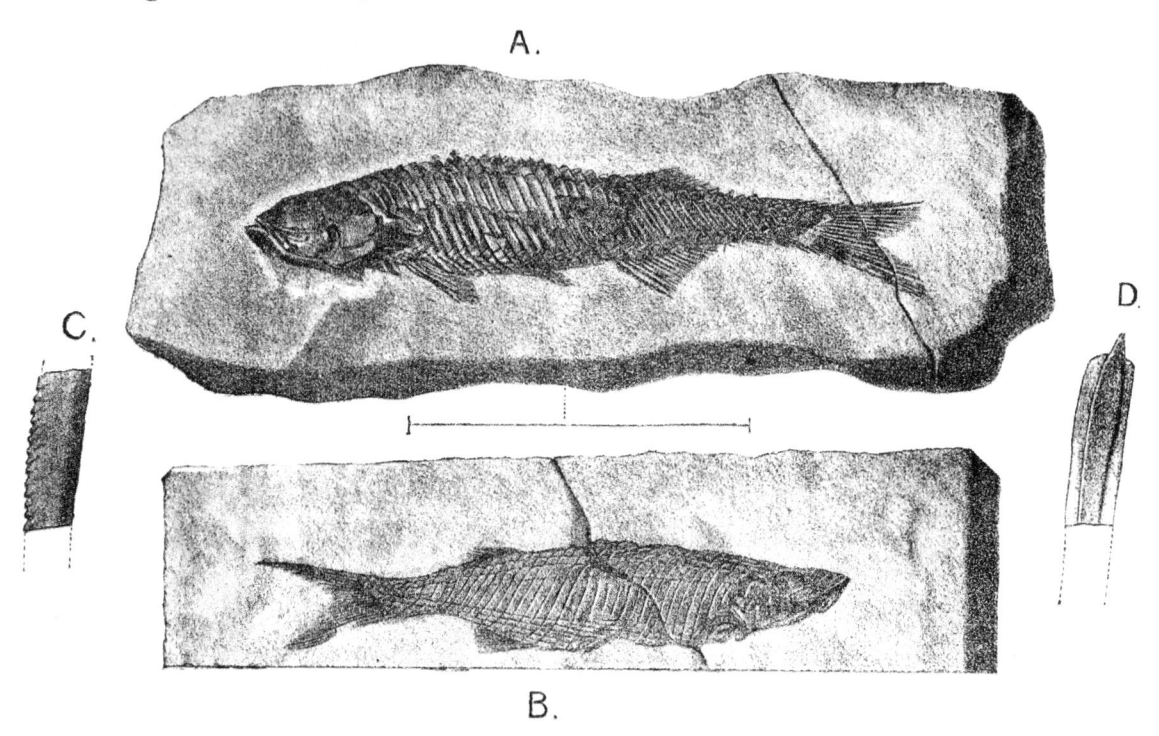

Fig. 35.—A. *Pleuropholis attenuata*, Egerton; type specimen, about twice nat. size.—Middle Purbeck Beds; Apsel Lane, Sutton Mandeville, Wiltshire. B. *Pleuropholis longicauda*, Egerton; type specimen, nat. size.—Middle Purbeck Beds; Swanage, Dorset. C, D. *Pleuropholis serrata*, Egerton; portions of flank-scales, outer (C) and inner (D) views, enlarged.—Purbeck Beds; Hartwell, near Aylesbury, Buckinghamshire. After Egerton.

1. Pleuropholis attenuata, Egerton. Text-figure 35A.

1854. *Pleuropholis attenuatus*, J. Morris (*ex* Egerton, MS.), Catal. Brit. Foss., ed. 2, p. 339 (name only).
1858. *Pleuropholis attenuatus*, P. M. G. Egerton, Figs. and Descripts. Brit. Organic Remains (Mem. Geol. Surv.), dec. ix, no. 7, p. 1, pl. vii, fig. 1.
1895. *Pleuropholis attenuata*, A. S. Woodward, Catal. Foss. Fishes, Brit. Mus., pt. iii, p. 483.

Type.—Imperfect fish; apparently lost.

Specific Characters.—The type species [known only by one specimen 5 cm. in length]. Length of head with opercular apparatus somewhat exceeding maximum depth of trunk and contained about five times in total length of fish; caudal pedicle slender, about half as deep as deepest flank-scale. Opercular bones smooth. Pelvic fins arising midway between the pectorals and the anal; dorsal fin, with 10 rays, arising opposite the origin of the anal fin, which comprises at least 12 rays, and arises nearly midway between the pectoral and caudal fins. Scales smooth, not serrated.

Remarks.—This species is still known only by the type specimen discovered by Mr. H. W. Bristow, which is now missing. The above definition is based on Egerton's description and figure (reproduced here in Text-fig. 35A).

Horizon and Locality.—Middle Purbeck Beds: Apsel Lane, Sutton Mandeville, Wiltshire.

2. **Pleuropholis formosa**, sp. nov. Plate XXII, fig. 8; Plate XXIII, figs. 8—11; Text-figure 36.

Type.—Nearly complete fish; British Museum.

Specific Characters.—A slender, regularly fusiform species, attaining a length of about 7 cm. Length of head with opercular apparatus about equalling maximum depth of trunk, and contained from five to six times in total length of fish; caudal pedicle about two-thirds as deep as deepest flank-scale. Opercular bones smooth. Pelvic fins arising slightly nearer to the anal than to the pectorals; dorsal fin, with 9 or 10 rays, arising just behind the origin of the anal, which is somewhat larger, with 11 or 12 rays, of which the length of the foremost is somewhat less than the depth of the trunk at its insertion. Scales smooth, not serrated.

Description of Specimens.—The type specimen, though fractured, seems to exhibit the general shape of the fish, with the complete caudal fin and enough remains of the other fins to indicate their position and proportions (Pl. XXIII, fig. 8). Parts of five smaller specimens on the same slab of limestone show various additional details; and several distorted examples obtained by the Rev. W. R. Andrews and others from the same horizon and locality confirm and extend our knowledge of the osteology of the species.

All the external bones of the head and opercular apparatus seem to have been smooth, covered with shining ganoine, and marked only by the occasional ridges and pores of the slime-canals. The cranial roof slopes gently downwards and forwards, without any marked bend in the frontal region; but in the basicranial axis the parasphenoid, as seen in side view crossing the orbit (Pl. XXIII, fig. 9, *pas.*), is much arched, inclining downwards and forwards as it leaves the pro-otic region, and then turning upwards to the short ethmoid region. The orbit is very large, and there is a delicate ossification in the sclerotic. The single series of cheek-plates round the orbit is narrow, and marked only by the large slime-canal, which traverses the orbital margin as usual. The mandibular suspensorium is much curved forwards, so that the quadrate articulation is beneath the anterior half of the orbit, and the gape of the mouth is very small. The entopterygoid (Pl. XXIII, fig. 9, *enpt.*) is a relatively large delicate lamina of bone, while the ectopterygoid (*ecpt.*) forms a stout bar at its lower border in front of the quadrate. The maxilla and premaxilla are unknown. The mandibular ramus (*md.*) is especially short and deep, and its dentary portion must have been comparatively small. There seem to be traces of minute conical teeth near the front of both jaws.

The operculum, displayed partly from within, partly as an impression in Pl. XXIII, fig. 9 (*op.*), is not less than two-thirds as wide as deep, and narrowed towards the upper end. Its outer face is somewhat convex, and quite smooth, though marked by a feeble waviness concentric with the lines of growth. The suboperculum and interoperculum are comparatively small: the former is much wider than deep, while the latter seems to taper upwards into a point between the operculum and preoperculum. The preoperculum, seen from within in Pl. XXIII, fig. 9 (*pop.*), is sharply bent at the expanded angle, the tapering upper and lower

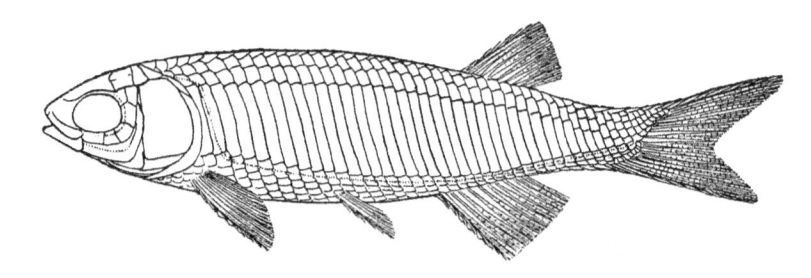

Fig. 36.—*Pleuropholis formosa*, sp. nov.; restoration, somewhat enlarged.—Lower Purbeck Beds; Teffont, Wiltshire.

limbs being nearly equal in size and at right-angles to each other. It is traversed by a conspicuous slime-canal, from which two or three straight branches radiate backwards at the angle. There are traces of large branchiostegal rays below the interoperculum.

The axial skeleton of the trunk is usually obscured by the thick squamation, but delicate cylindrical vertebral centra are observable in some broken specimens discovered by the Rev. W. R. Andrews (B.M. no. P. 9851).

Each of the pectoral fins comprises a dozen rays, of which the length of the longest about equals that of the head without operculum. Near its base the fringing fulcra are normal, but in its distal half they are more widely spaced and each bears an ovate expansion at its apex (Pl. XXIII, fig. 10). The pelvic fins are only about two-thirds as large as the pectorals, with not less than 6 rays and normal slender fulcra. The dorsal and anal fins are about equally elevated, with 9 and 11 rays respectively, of which the length of the foremost is less than the depth of the trunk at its insertion. As shown when the squamation is removed (Pl. XXIII, fig. 11), each ray is separated from its corresponding support by a short intercalated rod, as in the existing *Amia*. All the rays bifurcate once or twice distally, with wide articulations. As in the type specimen, the fulcra are usually slender and normal, becoming very small before disappearing distally; but in one specimen (B.M. no. P. 6304) the fulcra of the anal fin bear expansions like those already described on the pectoral fulcra. The caudal fin is nearly complete in the type specimen, displaying its 16 rays, of which 4 in the middle are well spaced. Its deeply overlapping fulcra are very slender, and become minute distally before they disappear. The upper caudal lobe of the body is prominent.

As shown by the type specimen, all the scales are smooth, without posterior serrations, and they are arranged in about 40 regular transverse series. All, except those near the end of the tail, are strengthened within by a broad vertical ridge, and united by a peg-and-socket articulation. The deepened flank-scale occurs in more than 30 of these series before it begins to subdivide on the caudal pedicle, being from four to five times as deep as its complete width in the front part of the abdominal region, and gradually becoming not much more than twice as deep as wide before subdivision on the caudal region. Each scale exhibits a slightly sigmoidal bend, and its lower end is truncated, while its upper end tapers a little as it curves forwards. Above each large flank-scale there are two other scales, of which the lower at least is deeper than wide; and there seem to be indications of the large flattened dorsal ridge-scales, such as are well seen in an example of another species of *Pleuropholis* from the Lithographic Stone of Cirin, Ain, France, in the British Museum (no. P. 4691a). The foremost upper scale is traversed by the slime-canal of the lateral line, which then passes down the second deepened flank-scale, and continues its course along a row of nearly square scales adjacent to the lower ends of the deepened flank-scales. Usually in the fossils the tubular excavation of the lateral line is exposed, but when a scale is perfect its only mark is a faint ridge. In the abdominal region there are four scales, wider than deep, beneath that of the lateral line, the lowest apparently forming the ventral ridge. At the base of the anal fin one row occurs beneath the lateral line, while in the caudal region beyond there seem to be three rows. On the caudal pedicle the deepened flank-scale is divided by a transverse suture into two, then into three scales, and the terminal scales are very small and rhombic. The lateral line ends abruptly at the insertion of the lowest of the four spaced caudal fin-rays. The acutely pointed and deeply overlapping caudal ridge-scales which pass into the fulcra of the caudal fin above and below are only moderately enlarged.

Immature Fish (Pl. XXII, fig. 8).—The immature *Pleuropholis*, only about 3 cm. in length, shown enlarged in Pl. XXII, fig. 8, has a relatively large head; but it may well be the fry of the species now described. It still lacks all the scales except the middle portion of those of the flank, and thus displays the internal skeleton. The remains of the head are much broken by crushing, but the curved parasphenoid is distinct in side view, and the relatively small mandible is observable. The course of the persistent notochord is indicated by a vacant space, but the neural and hæmal arches are well calcified. In the abdominal region, the right and left halves of each neural arch are separate both from each other and from the neural spine, which is comparatively short; the ribs, which do not reach the ventral border, are very slender, but attached to rather stout triangular ossifications abutting on the notochord. In the caudal region both the neural and hæmal arches are fused with their recurved spines, which are also small. Beneath the upturned end of the notochord the hæmal arches supporting the caudal fin are much enlarged and thickened; and in the upper caudal lobe there are traces of a separate series of short rods below the fulcral ridge-scales. Fulcra are not clear on any of the fins except the caudal, but the rays and their supports are well calcified. Six divided rays and one short simple anterior ray are seen in one of the pelvic fins. Nine and eleven divided rays respectively can be counted in the dorsal (fig. 8a) and anal fins, besides a few short simple rays in front, each ray (except perhaps anteriorly) being separated from its corresponding support by a short intercalated rod, as in *Amia*. It may be added that there are traces of ganoine on some of the rudimentary flank-scales. White granular material, probably phosphatic, in the body-cavity may represent food.

Horizon and Locality.—Lower Purbeck Beds: Teffont, Vale of Wardour, Wiltshire.

3. Pleuropholis crassicauda, Egerton. Plate XXIII, figs. 12, 13.

1858. *Pleuropholis crassicaudus*, P. M. G. Egerton, Figs. and Descripts. Brit. Organic Remains (Mem. Geol. Surv.), dec. ix, no. 7, p. 3, pl. vii, fig. 2.
1895. *Pleuropholis crassicaudata*, A. S. Woodward, Catal. Foss. Fishes, Brit. Mus., pt. iii, p. 484, pl. xiv, fig. 5.

Type.—Portion of fish; British Museum.

Specific Characters.—As *P. formosa*, so far as known, but the trunk relatively deeper, its maximum depth a little exceeding the length of the head with opercular apparatus.

Description of Specimens.—The type specimen in the P. B. Brodie Collection (Pl. XXIII, fig. 12) is too imperfect for specific determination, but it evidently represents the same species as a more nearly complete specimen from the same horizon and locality shown enlarged in Pl. XXIII, fig. 13. The latter may therefore be used for the specific diagnosis, as above.

The type specimen (Pl. XXIII, fig. 12) seems to show the complete length of the head, with indications of the very large orbit, stained black, the short ethmoid region, and the narrow postorbitals traversed by a large slime-canal. The upper limb of the preoperculum, also with a conspicuous slime-canal, is distinct; and the nearly complete operculum, about two-thirds as wide as deep, is seen to be smooth, but marked with a slight waviness concentric with the lines of growth.

The second specimen (fig. 13), of which the jaws, shown mainly in impression, are somewhat displaced forwards, exhibits also traces of an ossified sclerotic.

Only fragments of the fins are preserved, but the places of origin of all except the dorsal are seen in the second specimen, where the extent of the forked caudal is also traceable. Stout deeply overlapping fulcra fringe the pelvic, anal, and caudal fins, those at the origin of the lower lobe of the caudal fin in the type specimen being displaced and especially conspicuous. The distantly articulated rays in the lower caudal lobe are especially well enamelled.

The scales must have been in about 40 transverse series. Some of the deepened flank-scales in the type specimen exhibit an irregular waviness of the outer face, doubtless following lines of growth, while those of the second specimen are exposed from within, showing the broad vertical ridge on their inner face, and the peg-and-socket articulation. The caudal scales preserved in the type specimen are clearly not serrated, while those in the second specimen, though bearing the internal ridge, lack the peg-and-socket articulation. As shown in Pl. XXIII, fig. 13, the lateral line curves down the second deepened flank-scale to traverse the usual ventral row of nearly equilateral scales; as shown on the tail of the type specimen, its course is marked by a slight smooth ridge on the outer face.

Horizon and Locality.—Middle Purbeck Beds: Durdlestone Bay, Swanage, Dorset.

4. **Pleuropholis longicauda**, Egerton. Plate XXIV, figs. 1, 2; Text-figure 35B.

1858. *Pleuropholis longicaudus*, P. M. G. Egerton, Figs. and Descripts. Brit. Organic Remains (Mem. Geol. Surv.), dec. ix, no. 7, p. 3, pl. vii, fig. 4.
1895. *Pleuropholis longicaudata*, A. S. Woodward, Catal. Foss. Fishes, Brit. Mus., pt. iii, p. 483.

Type.—Imperfect fish; apparently lost.

Specific Characters.—Imperfectly known, but attaining a length of about 9 cm., and the head with opercular apparatus occupying somewhat more than one-sixth of this length. Probably distinguished from all other known species by the depression of the head, which seems to have been scarcely more than half as deep as the middle part of the trunk. Scales smooth, not serrated; the rhombic dorsal and ventral scales often produced to a sharp point at their postero-inferior angle.

Description of Specimens.—The type specimen (Text-fig. 35B, p. 114), which was in Mr. W. R. Brodie's Collection, cannot now be traced, and the description and figure given by Egerton are insufficient to diagnose the species. One specimen in the British Museum, however, from the Middle Purbeck Beds of Swanage, is labelled "*Pleuropholis longicaudus*" by Egerton, and two other specimens from the same formation and locality also seem to belong to this species. It may, therefore, probably be defined as above.

The head seems to be crushed upwards in the type specimen. Its comparatively small size is best shown in the original of Pl. XXIV, fig. 1, but the bones are so much broken by crushing that their outlines are rather obscured. The relatively large orbit is bounded behind, below, and in front by the single series of smooth cheek-plates, which are traversed by the usual circumorbital slime-canal. The gape of the mouth is evidently small. The opercular bones, partly seen also in Pl. XXIV, fig. 2, are smooth. The large angularly-bent preoperculum is marked only by the groove of the slime-canal. The operculum is about two-thirds as wide as deep, and the suboperculum is relatively small. The interoperculum seems to be triangular in shape and deeper than wide, its apex rising between the lower end of the operculum and the preoperculum.

Delicate ring-vertebræ and short slender ribs are exposed by the removal of the scales in the specimen labelled by Egerton (B.M. no. P. 1101).

The bases of about nine very stout rays are seen spread out in the pectoral fin of Pl. XXIV, fig. 1; and the foremost pectoral ray in fig. 2 bears a regular series of very slender fulcra. The pelvic fins, which are shown in two specimens to arise midway between the pectorals and the anal, also consist of unusually stout rays. The dorsal and anal fins arise directly opposite each other, but are incompletely known.

The scales are clearly all smooth without any posterior serrations. The depth of the deepest flank-scale, above the pelvic fins, slightly exceeds four times its complete width (fig. 2a). The rhombic dorsal and ventral scales are often produced into a sharp point at their postero-inferior angle, and both the dorsal and ventral ridge-scales are sharply pointed behind (figs. 1a, 2a). Two rows of scales intervene as usual between the principal flank-scales and the dorsal ridge-series, the lower being slightly the deeper; while five small scales occur below each principal flank-scale as far back as the pelvic fins, the uppermost being traversed by the lateral line. The ventral scales become fewer behind, as already described in *P. formosa* (p. 117), and the stout ventral ridge-scales are comparatively small.

Horizon and Locality.—Middle Purbeck Beds: Swanage, Dorset.

5. **Pleuropholis serrata,** Egerton. Text-figure 35c, D.

1858. *Pleuropholis serratus,* P. M. G. Egerton, Figs. and Descripts. Brit. Organic Remains (Mem. Geol. Surv.), dec. ix, no. 7, p. 5, pl. vii, figs. 5—9.
1895. *Pleuropholis serrata,* A. S. Woodward, Catal. Foss. Fishes, Brit. Mus., pt. iii, p. 487.

Type.—Imperfect fish; collection of late Dr. John Lee, Hartwell.

Specific Characters.—Imperfectly known, but a comparatively stout fish attaining a length of about 10 cm. Head and opercular bones smooth. Scales also smooth, but those of the deepened flank-series with fine oblique serrations on the hinder border.

Description of Specimens.—All the known examples of this species are much crushed and broken, but it is easily distinguished from the other Purbeckian species by its less elegant proportions and by the serration of the principal flank-scales. Both the head and opercular bones are shown to have been smooth. The vertebral centra are preserved as delicate cylinders, with very little constriction. The fulcra on the anal fin are very slender. The fine oblique serration of the principal flank-scales has already been described and figured by Egerton, who also notes the broad thickened band on the inner face connected with the peg-and-socket articulation (Text-fig. 35c, D, p. 114). There is an occasional opening for slime-apparatus in these scales, but the lateral line as usual clearly passes along the row of small scales immediately below. The dorsal and ventral scales do not appear to be serrated.

Horizon and Localities.—Purbeck Beds: Hartwell and Bishopstone, Buckinghamshire.

Family OLIGOPLEURIDÆ.

These are the latest primitive teleosteans which retain fulcra on all the fins. They occur chiefly in the Lower Kimmeridgian (Lithographic Stone) of France and Germany, though they also range through Cretaceous formations. Some Wealden and Purbeck fossils have been wrongly referred to *Oligopleurus* itself on imperfect evidence (see p. 129), but a maxilla (Pl. XVII, fig. 8), from the Middle Purbeck of Swanage, may perhaps belong to the allied genus *Œonoscopus*, Costa. This bone much resembles the maxilla of *Megalurus* and *Amiopsis*, but it agrees still more closely with the same bone in two specimens of *Œonoscopus cyprinoides*, from the Lithographic Stone of Bavaria, in the British Museum (as already mentioned in Geol. Mag. [4], vol. ii, 1895, p. 151, pl. vii, fig. 9). It is much laterally compressed, and more than twice as deep behind as in front; its hinder margin is slightly excavated, while its anterior end is produced into a stout incurved portion for articulation with the palatine. Its outer face is almost smooth, being very faintly rugose, and the oral margin is only slightly sinuous. The teeth do not vary much in size, and are small, but stout and conical, with the blunt enamelled apex turned somewhat inwards. They are hollow and smooth, and closely though irregularly arranged. Some teeth are broken away from the gaps observed in the series.

Family LEPTOLEPIDÆ.

Genus **LEPTOLEPIS**, Agassiz.

Leptolepis, L. Agassiz, Neues Jahrb. für Min., etc., 1832, p. 146.
Oxygonius, L. Agassiz, in Brodie's Fossil Insects, 1845, p. 16.
Tharsis, C. G. Giebel, Fauna der Vorwelt, Fische, 1848, p. 145.
Sarginites, O. G. Costa, Atti Accad. Pontan., vol. v, 1850, p. 285.
Megastoma, O. G. Costa, loc. cit., 1850, p. 287.

122 WEALDEN AND PURBECK FOSSIL FISHES.

Generic Characters.—Head large and teeth minute; sclerotic ossified. Maxilla arched, with a slightly convex tooth-bearing border; mandible prominent, and dentary rising sharply into a thickened, obtuse elevation near its anterior end; preoperculum broad mesially, with a large inferior limb, marked with radiating ridges; subopercBulum large, but smaller than the trapezoidal operculum, from which it is divided by an oblique suture. Vertebral centra in the form of much-constricted cylinders, with little or no secondary ossification. Pelvic fins relatively large; dorsal fin about as long as deep, opposed to the pelvic pair or to the space between the latter and the anal; anal fin small, not much extended; caudal fin deeply forked. Scales completely covering the trunk; no enlarged or thickened ridge-scales.

Type Species.—*Leptolepis bronni* (L. Agassiz, Poiss. Foss., vol. ii, pt. i, 1833,

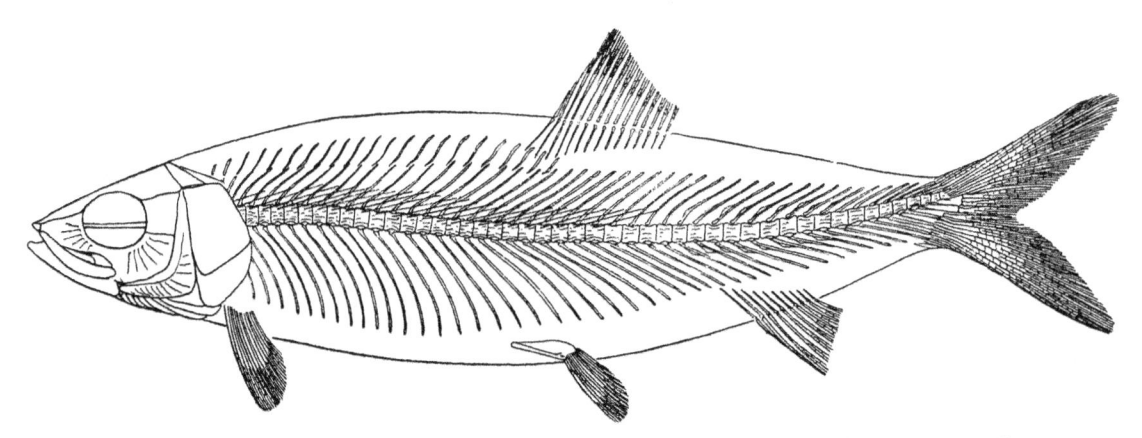

FIG. 37.—*Leptolepis dubius* (Blainville); restoration of skeleton, scales omitted, reduced in size.—Lower Kimmeridgian (Lithographic Stone); Bavaria. From British Museum Catalogue of Fossil Fishes.

p. 13; pt. ii, 1844, pp. 133, 294), from the Upper Lias of Würtemberg, Bavaria, France, and England.

Remarks.—The most characteristic bone of *Leptolepis* is the dentary of the mandible (Pl. XXIII, fig. 7). The genus has a very wide distribution both in time and space, ranging from the Upper Lias to the Lower Cretaceous in Europe, and from Spitzbergen in the north to Australia in the south. The later species as a rule (*e. g. Leptolepis dubius*, Text-fig. 37) exhibit more secondary ossification in the vertebræ than the earlier species; but a Kimmeridgian form from King Charles Land (*L. nathorsti*, A. S. Woodward, Bihang K. Svensk. Vet.-Akad. Handl., vol. xxv, sect. iv, no. 5, 1900, p. 4, figs. 2—11) and the Purbeckian species described below seem to have primitive vertebral centra like those of *L. bronni*.

1. **Leptolepis brodiei**, Agassiz. Plate XXIII, figs. 1—6.

1845. *Leptolepis brodiei*, L. Agassiz, in P. B. Brodie, Foss. Insects, p. 15, pl. i, figs. 1, 3.
1845. *Leptolepis nanus*, P. M. G. Egerton, in P. B. Brodie, *op. cit.*, p. 15, pl. i, fig. 5.

1845. *Oxygonius tenuis*, L. Agassiz, in P. B. Brodie, *op. cit.*, p. 16, pl. i, fig. 4.
1850. *Sarginites pygmæus*, O. G. Costa, Atti Accad. Pontan., vol. v, p. 285, pl. vi, figs. 6—8.
1850. *Megastoma apenninum*, O. G. Costa, *loc. cit.*, vol. v, p. 287, pl. vi, figs. 9, 10.
1853. *Sarginites pygmæus*, O. G. Costa, *loc. cit.*, vol. vii, p. 7, pl. i, fig. 4.
1853. *Megastoma apenninum*, O. G. Costa, *loc. cit.*, vol. vii, p. 8, pl. i, fig. 3.
1879. *Leptolepis neocomiensis*, F. Bassani, Verhandl. k. k. geol. Reichsanst., p. 164.
1882. *Leptolepis neocomiensis*, F. Bassani, Denkschr. k. Akad. Wiss. Wien, math.-naturw. Cl., vol. xlv, p. 204, pl. ii, figs. 1—5.
1895. *Leptolepis brodiei*, A. S. Woodward, Geol. Mag. [4], vol. ii, p. 150, pl. vii, figs. 5, 6; also Catal. Foss. Fishes Brit. Mus., pt. iii, p. 515.
1912. *Leptolepis brodiei*, F. Bassani and G. D'Erasmo, Mem. Soc. Ital. Sci., XL [3], vol. xvii, p. 229, pl. iv, fig. 6.
1915. *Leptolepis brodiei*, F. Bassani and G. D'Erasmo, Palæont. Italica, vol. xxi, p. 12, pl. i, figs. 4—6.

Type.—Imperfect fish; British Museum.

Specific Characters.—A small species, attaining a length of about 5 cm., but usually smaller. Length of head with opercular apparatus exceeding the maximum depth of the trunk, and contained about four times in the total length of the fish; width of caudal pedicle at least half the maximum depth of the abdominal region. Vertebræ from 40 to 45 in number, about half being caudal; the centra smooth; the neural and hæmal spines in the caudal region nearly straight. Pelvic fins arising about midway between the pectoral and anal fins, opposite to the anterior half of the dorsal, which has 10 or 11 bifurcated rays besides 2 or 3 shorter simple rays in front; anal fin with about 7 rays, arising nearer to the caudal than to the pelvic fins.

Description of Specimens.—The type specimen, which is shown enlarged in Pl. XXIII, fig. 3, is a little distorted by the crushing upwards of the ventral border of the hinder abdominal region, while its caudal region is imperfect and lacks the caudal fin. The clavicles of the fish may also be slightly crushed backwards. It shows, however, most of the characteristic features of the species. The second specimen figured by Brodie is similarly crushed, and is important as displaying the distinctive shape of the dentary bone (Pl. XXIII, fig. 4). Other specimens in the Brodie Collection, and still finer examples discovered by the Rev. W. R. Andrews, nearly complete our knowledge of the species. Two of these seem to show the true shape of the trunk and the proportions of the caudal fin (Pl. XXIII, figs. 5, 6).

All these fishes have the appearance of immaturity, with a relatively large orbit, which is often marked by a black stain. The head and opercular bones are smooth, and the slime-canals are relatively large, as shown especially by the frontals when exposed from below (B.M. no. P. 7635). The slender parasphenoid is nearly straight, but curves upwards a little both in front and behind. The mandibular suspensorium is inclined much forwards, so that the articulation is beneath the middle of the orbit. In one specimen discovered by Rev. W. R. Andrews (B.M. no. P. 6305) minute teeth seem to occur on the slender arched maxilla, which has a thickened upper border. In three specimens the ceratohyal is shown as a simple hour-glass-shaped bone, without any ossified filament connecting its ends. The preoperculum, as usual, is sharply bent at its angle, with a relatively large lower limb; and there seem to be two or three ridges radiating backwards from the slime-canal on the slight expansion at the angle. There are not less than 10 branchiostegal rays, of which the foremost are relatively small, slender, and spaced.

The exact number of the vertebræ is uncertain, but there are about 22 in the abdominal region and at least 20 in the tail. About four are comprised within the branchial region, and there are five or six in the upturned end which supports the caudal fin. As shown both by the type specimen and by others, the centra are delicate constricted cylinders pierced by a considerable remnant of the notochord, and their outer surface is smooth, not strengthened by longitudinal ridges. The neural arches are longest in the anterior half of the abdominal region, where they support loosely-apposed stout neural spines as far back as the origin of the dorsal fin (Pl. XXIII, fig. 3). Beneath the dorsal fin the neural arches are much shortened, without separate neural spines, and in the caudal region they are also short, sharply inclined backwards, and symmetrical with the hæmal arches. The slender arched ribs extend nearly to the ventral border, and the expanded hæmal arches supporting the caudal fin are all separate (Pl. XXIII, fig. 5). Intermuscular bones, very delicate, are seen only lying across the neural spines in the abdominal region.

As shown in both the specimens figured by Brodie (re-figured in Pl. XXIII, figs. 3, 4), the pelvic fins are not much less elevated than the pectorals. They comprise 7 or 8 rays, and seem to be inserted opposite the front half or the middle of the dorsal, their position varying in the fossils according to the manner in which they are crushed. The dorsal fin is especially well seen in the type specimen, where its maximum depth somewhat exceeds that of the trunk (fig. 3). Its foremost three rays are short and simple, gradually increasing in length; next there are 10 or 11 rays which bifurcate, and are distantly articulated in their distal half and slightly diminish in size backwards. Of the fin-supports the foremost two or three are fused together at an acute angle, while above the others there are distinct traces of the short intercalary pieces which are well known in *Amia* and several fossil Amioids. The remote anal fin, best seen in Pl. XXIII, fig. 6, is small, and comprises only about 7 rays, rapidly shortening backwards. Its rays also bifurcate and are distantly articulated in their distal half. The forked caudal fin is especially powerful (Pl. XXIII, figs. 5, 6), with distantly-articulated bifurcating rays, of which the five or six shortest in the middle of the fork are spaced. Stout sigmoidally-bent fulcral rays are seen at its base both above and below (fig. 6).

The thin cycloidal scales are often seen in the fossils, marked only by the concentric lines of growth; but no enlarged or thickened ridge-scales have been observed.

It may be added that in one of the specimens figured (Pl. XXIII, fig. 5, c.) the intestine is conspicuous, filled with white fæcal matter.

Immature Fishes.—The type specimen of the so-called *Leptolepis nanus*, which is re-drawn enlarged in Pl. XXIII, fig. 2, is obviously an immature example of *Leptolepis* much shortened by distortion, and there is nothing to distinguish it from *L. brodiei*. Below and in front of the large orbit, which is stained black, the characteristic dentary bone is conspicuous. The telescoped vertebræ irregularly overlap, but the crushed dorsal and pelvic fins remain almost in their natural relative positions. The base of the caudal fin is widened by the distortion.

There can also be no doubt that the immature fishes named *Oxygonius tenuis* by Agassiz are very young individuals of *Leptolepis*, almost certainly of *L. brodiei*. Their relatively large and delicately ossified vertebral column is a well-known mark of immaturity.[1] The type specimen, which is re-drawn enlarged in Pl. XXIII, fig. 1, is made a little slender by crushing, but not otherwise distorted. Its skull is missing, but the orbit is marked by a black stain, and the characteristic jaws are seen. The relative width of the opercular apparatus is also clear. The feebly ossified vertebral centra, in close series, are much deeper than long, not constricted, and apparently about 40 in number. Slight traces of the pectoral fins and the small remote anal fin are preserved, while the dorsal and pelvic fins are nearly complete, though depressed by crushing. The large forked caudal fin is also well displayed. Among other specimens labelled by Brodie, one in a group in the British Museum (no. P. 7634) exhibits very clearly the ascending process of the dentary bone, and others show that the fins must have been as in *L. brodiei*.

Horizon and Locality.—Lower Purbeck Beds: Vale of Wardour, Wiltshire.

Specifically indeterminable fragments of *Leptolepis* have also been found in the Wealden. Among them may be specially noted the dentary bone of the mandible shown enlarged in Pl. XXIII, fig. 7.

Genus ÆTHALION, Münster.

Æthalion, G. von Münster, Neues Jahrb. f. Min., etc., 1842, p. 41.

Generic Characters.—As *Leptolepis*, but dentary bone of mandible gradually deepening from the symphysis backwards without any marked thickening, and vertebral centra much strengthened by secondary ossification in fine longitudinal ridges.

Type Species.—*Æthalion knorri* (*Clupea knorri*, H. D. de Blainville, Nouv. Dict. d'Hist. Nat., vol. xxvii, 1818, p. 331), from the Lower Kimmeridgian (Lithographic Stone) of Bavaria.

---

[1] W. C. McIntosh and A. T. Masterman, The Life-Histories of the British Marine Food-Fishes (1897), p. 415.

Remarks.—This genus occurs chiefly in the Lithographic Stone of Bavaria, France, and Northern Spain (*A. vidali*, H. E. Sauvage, Mem. R. Acad. Cienc. Barcelona [3], vol. iv, no. 35, 1903, p. 13, pl. ii, fig. 2). One species has been identified from the Wealden of Belgium (*A. robustus*, R. H. Traquair, Poiss. Weald. Bernissart, Mém. Mus. Roy. Hist. Nat. Belg., vol. v, 1911, p. 50, pl. xi), and the same or a closely allied form is recorded from the Lower Cretaceous of Castellamare, Naples (F. Bassani and G. D'Erasmo, Mem. Soc. Ital. Sci. XL [3], vol. xvii, 1912, p. 234, pl. iii, fig. 3, pl. vi, figs. 1, 2, text-figs. 13—15; also G. D'Erasmo, Palæont. Italica, vol. xxi, 1915, p. 15, pl. i, fig. 8, text-fig. 22). Dr. Traquair's

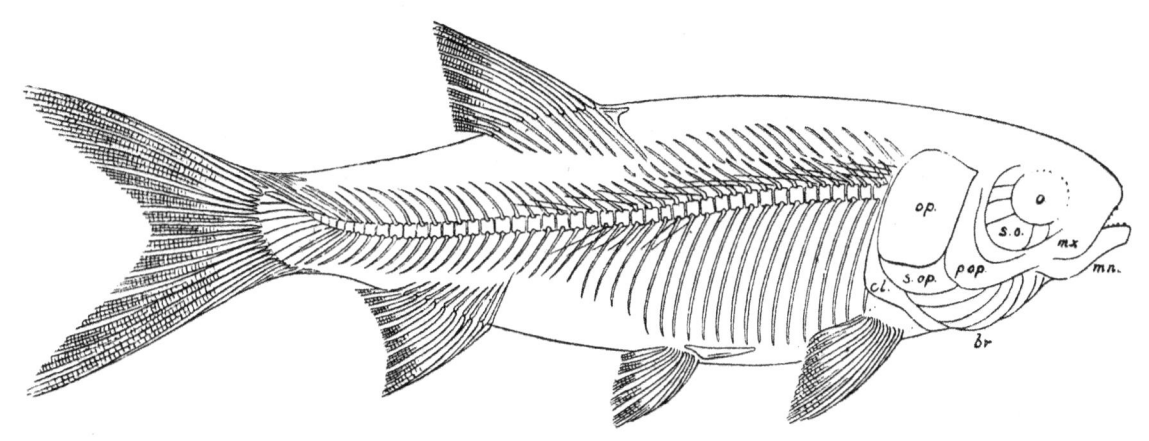

Fig. 38.—*Æthalion robustus*, Traquair; restoration of skeleton, scales omitted, reduced in size.—Wealden; Bernissart, Belgium. After R. H. Traquair. *br.*, branchiostegal rays; *cl.*, clavicle; *mn.*, mandible; *mx.*, maxilla; *o.*, orbit; *op.*, operculum; *pop.*, preoperculum; *s.o.*, cheek-plates; *s.op.*, subopercnlum.

sketch of the restored skeleton of the Wealden *A. robustus* is reproduced in Text-fig. 38.

1. **Æthalion valdensis**, A. S. Woodward. Text-figure 39.

1907. *Leptolepis valdensis*, A. S. Woodward, Ann. Mag. Nat. Hist. [7], vol. xx, p. 93, pl. i.

Type.—Imperfect fish; British Museum.

Specific Characters.—A stout species attaining a length of about 40 cm. Length of head with opercular apparatus exceeding the maximum depth of the trunk, which equals about one-third the length from the pectoral arch to the base of the caudal fin, and would probably equal about one-fifth of the total length of the fish. Vertebræ 60. Pelvic fins arising midway between the pectoral and anal fins, opposite the front half of the dorsal, which arises much nearer to the occiput than to the caudal fin and comprises 18 to 20 rays; anal fin with about 14 rays, arising slightly nearer to the caudal than to the pelvic pair. Scales rather large and very deeply overlapping, some feebly crimped.

Description of Specimen.—The type specimen (Text-fig. 39) is still unique,

ÆTHALION. 127

but exhibits most of the characters of the species in the counterpart halves of a slab of clay. The mandibular suspensorium is clearly inclined forwards, so that the articulation of the lower jaw must have been directly beneath the hinder part of the orbit. The hyomandibular bone (*hm.*) bears a rather long process for the suspension of the operculum (*op.*), which is too imperfect to show its shape completely. The angle of the preoperculum (*pop.*) is much expanded, and its tapering ascending limb is upright. The suboperculum (*sop.*) must have been about four times as broad as deep. Fifteen branchiostegal rays (*br.*) can be counted, the upper seven being expanded and in close series, the lower eight being

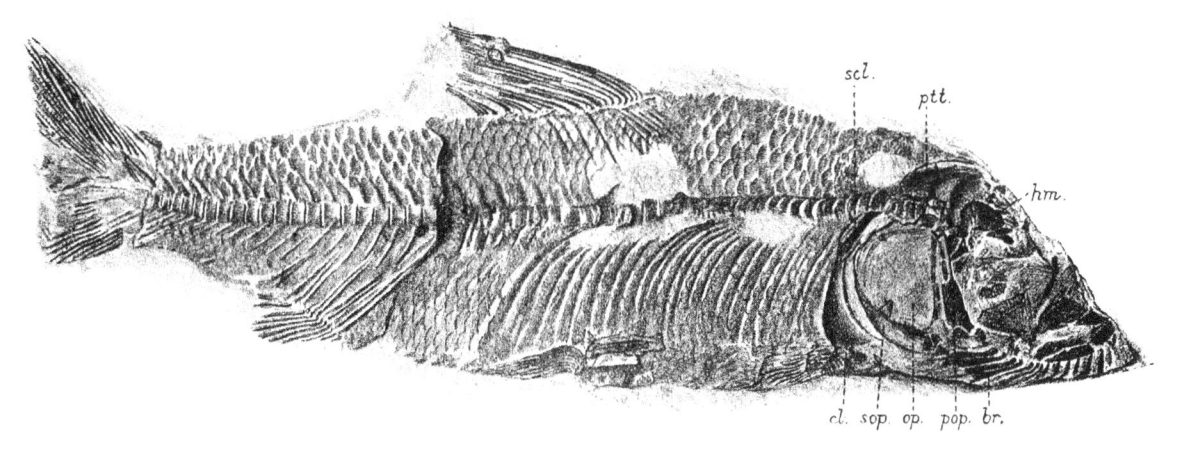

Fig. 39.—*Æthalion valdensis*, A. S. Woodward; type specimen, nearly one-half nat. size.—Weald Clay; Southwater, Sussex. British Museum, no. P. 10440. *br.*, branchiostegal rays; *cl.*, clavicle; *hm.*, hyomandibular; *op.*, operculum; *pop.*, preoperculum; *ptt.*, post-temporal; *scl.*, supraclavicle; *sop.*, suboperculum.

narrower bars and more widely spaced. All the opercular apparatus is smooth, not ornamented. The total number of vertebræ is about 60, half being in the abdominal region. The centra are about as long as deep in the anterior part of the caudal region, but are somewhat shorter than deep both in the abdominal and in the hinder part of the caudal region. They are well ossified, and their primitive double cone is strengthened by secondary bone arranged in fine, close, longitudinal ridges. The ribs are stout, apparently borne on very short transverse processes, and extending to the ventral border of the fish. The fixed neural and hæmal arches in the caudal region are also stout and gently arched. The hinder end of the vertebral column turns only slightly upwards, and its hæmal arches are expanded without fusion. The intermuscular bones are much obscured by the scales in the fossil, but there are traces of them above the vertebral column in the abdominal region, and both above and below this column in the caudal region.

The post-temporal bone (*ptt.*) is a thick plate, almost triangular in shape, and the supraclavicle (*scl.*) is a deep and narrow bone. The clavicle (*cl.*), as shown in impression, is expanded into a large smooth plate above the pectoral fin, which is inserted close to the ventral border. When adpressed to the trunk this fin extends at least half-way to the insertion of the pelvic fins. The latter are smaller than the pectorals, and their supports are a pair of elongated thin laminæ, which meet in the middle line and are thickened along their outer borders. The dorsal fin arises well in front of the middle point between the occiput and the caudal fin, comprising 18 to 20 rays, of which the three foremost are short, undivided, closely pressed together, and gradually increase in length. The length of the fourth or longest dorsal fin-ray somewhat exceeds two-thirds of the depth of the trunk at its insertion. The anal fin resembles the dorsal, but is much smaller, and comprises only 13 or 14 rays. The remains of the caudal fin-rays show that they are comparatively stout and closely articulated. The fulcral scales at the base of the upper caudal lobe are especially stout, and are continued up the foremost ray as a short fringe.

The scales are relatively large, cycloid, and smooth, occasionally with feeble traces of a radiating pectination at the hinder border, but usually exhibiting structural lines, including wavy concentric markings. They are scarcely displaced in the fossil, and are seen to be deeply overlapping, with the exposed area narrow and deep. The lateral line is scarcely traceable, but seems to produce a slight depression along some scales in a series above the vertebral column.

Remarks.—The fossil thus described evidently belongs either to *Leptolepis* or to *Æthalion*, but the absence of the mandible leaves its reference to one or other of these two genera uncertain. It was originally assigned to *Leptolepis*, but the extent of the intermuscular bones and the close articulation of the caudal fin-rays are suggestive rather of *Æthalion*; its general resemblance to the species of *Æthalion* described by Traquair from the Wealden of Bernissart, Belgium, is indeed noteworthy. It may, therefore, best be placed provisionally in the latter genus. Its number of vertebræ exceeds that of all the known species of both *Leptolepis* and *Æthalion*, except *A. vidali* from the Upper Jurassic of Spain.

Horizon and Locality.—Weald Clay: Southwater, Sussex.

Genus PACHYTHRISSOPS, novum.

Parathrissops, C. R. Eastman (*non* Sauvage, 1875), Mem. Carnegie Mus. Pittsburgh, vol. vi (1914), p. 423.

Generic Characters.—Head as in *Leptolepis*, but the elevation of the dentary relatively broad and less thickened. Vertebral centra much strengthened by secondary ossification in fine longitudinal ridges, short and deep in the abdominal region, sometimes longer in the tail. Pelvic fins relatively large; dorsal and anal fins acuminate, the dorsal about as long as deep, arising opposite or just in advance of the origin of the anal fin, which is equally deep and usually more extended; caudal fin deeply forked. Scales very delicate, often not preserved in the fossils.

Type Species.—*Pachythrissops lævis*, from the English Purbeck Beds.

Remarks.—The two species defined below were originally referred in error to *Oligopleurus*, and are proved by the new specimens now described to belong to the Leptolepidæ. They seem to be generically identical with a fish from the Lithographic Stone of Bavaria named *Parathrissops furcatus* by Eastman (Mem. Carnegie Mus. Pittsburgh, vol. vi, 1914, p. 423, pl. lix, fig. 2), which is distinguished by the characters of its dorsal and anal fins. The generic name *Parathrissops*, however, is preoccupied by Sauvage (Bull. Soc. Sci. Yonne, vol. xlv, pt. ii, 1891, p. 37). The species also bear some resemblance to *Eurystethus brongniarti* (H. E. Sauvage, Bull. Soc. Géol. France [3], vol. vi, 1878, p. 629, pl. xiii, fig. 2), from the Kimmeridgian of Morestel, Ain, France, but this fish is too imperfectly known for precise comparison.

1. **Pachythrissops lævis**, sp. nov. Plate XXIV, figs. 3—5; Plate XXV, figs. 1—3.

1890. *Oligopleurus* (?) *vectensis*, A. S. Woodward, Proc. Zool. Soc., p. 346, pl. xxviii, figs. 2—4, pl. xxix, figs. 1,—3.

Type.—Immature fish; British Museum.

Specific Characters.—The type species, probably attaining a length of 50 cm., but usually smaller. Length of head with opercular apparatus much exceeding maximum depth, also much exceeding distance between paired fins, and occupying nearly one-quarter of total length of fish. Operculum smooth, but with slime-pit defined by two ridges on outer face at point of suspension; preoperculum with few radiating ridges near the angle. Vertebræ 35 in abdominal, nearly 30 in caudal region; the centra of the latter longer than the former, and marked by a sharp lateral ridge. Dorsal and anal fins arising directly opposite, the former with about 18, the latter with slightly more than 20 rays.

Description of Specimens.—The type specimen (Pl. XXV, fig. 1) is a small fish, which must probably be regarded as immature. It is shown in direct side view, apparently not distorted, but its vertebral axis is not well preserved, and the ribs are scarcely seen. All the known larger specimens are fragmentary, and one of these (Pl. XXIV, fig. 4) was wrongly referred to *Lepidotus minor* by L. Agassiz, Poiss. Foss., vol. ii, pt. i (1844), p. 269, pl. xxix c, fig. 12.

The head is much crushed and broken in the type specimen, but enough is clear both in this and other specimens to identify with the same species the roof of a skull shown in Pl. XXIV, fig. 3. Here the specimen is somewhat flattened by vertical crushing, but the bones are complete as far forwards as the front of the orbit. The occiput is well ossified, and its upper portion is formed by a supra-occipital in the middle with a pair of epiotics at the sides. The supraoccipital (*socc.*) occupies the middle third, and bears a vertical median crest on its hinder hollowed face. The epiotic (*epo.*) is very prominent on each side, triangular in shape when viewed from above, nearly as long as broad, and united in a finely-toothed suture with the next otic element in front. Its upper face, like that of the supraoccipital, is smooth, and must have been overlapped by the supratemporal. Each parietal (*pa.*) is slightly more than twice as long as broad, meeting its fellow in an almost straight median suture, and deeply interdigitating with the frontal anteriorly. The middle of its hinder half is raised into a coarsely rugose boss, in which the partly reticulating ridges tend to radiate from the transverse groove of the slime-canal. Its anterior half is nearly smooth. Each squamosal (*sq.*) is about as wide as the pair of parietals, and equally long. It unites in a slightly wavy suture with the corresponding parietal, and in small digitations with the frontal, while its outer border is somewhat concave. Its outer face is nearly smooth, with some irregular pitting and longitudinal grooving. The relatively large frontals (*fr.*), which also meet in a nearly straight median suture, are much narrowed between the relatively large orbits, and resemble the squamosals in the irregular grooving and pitting of the nearly smooth outer face. The postfrontals or sphenotics (*ptf.*) form a conspicuous pair of smooth, truncated prominences about as long as wide. There is a well-marked median depression in the cranial roof, beginning behind between the parietal bosses, and widening and deepening to the greatest extent between the hinder part of the orbits. The whole appearance of the bones suggests a considerable development of the slime-apparatus. The otic region and the basioccipital are well ossified, and the parasphenoid is comparatively stout, bearing an elongated patch of minute teeth where it underlies the hinder part of the orbit (seen in B.M. no. 21974). In some specimens there are broken traces of very thin cheek-plates. As shown in the type specimen, the mandibular suspensorium is arched forwards, so that the articulation for the mandible is beneath the middle of the orbit. The hyomandibular is a deep and much expanded thin lamina of bone, but is known only in a crushed specimen (B.M. no. P. 4535). It is strengthened by a vertical ridge, from which a short ridge proceeds at right-angles to an elongated articular process for the operculum. The triangular quadrate is cleft behind to accommodate a rod-shaped symplectic, and articulates above with a large triangular metapterygoid. The ectopterygoid (best seen in Pl. XXIV, fig. 4, *ecpt.*) is an arched, thin lamina of bone tapering forwards, and the entopterygoid seems to be still thinner. It is uncertain whether these elements bore teeth. As shown in the type specimen, the premaxilla is more extended backwards than usual in Amioids; and, as seen in another specimen (B.M. no. P. 4535), it bears a single series of small but stout, hollow, conical teeth. The maxilla, of which the middle portion is broken away in the type specimen, is relatively large and arched, with a flattened and nearly smooth outer face, and the convex oral border bearing minute teeth. Its posterior half is overlapped by two supramaxillæ, of which the posterior is the larger, and excavated in front for the anterior, which is elongate-triangular in shape. In the mandible, the dentary bone is relatively large, and its posterior end extends along the lower border almost as far as the position of the mandibular articulation. Though well shown in the type specimen, its complete shape is better seen in the original of Pl. XXV, fig. 2. The bone is truncated at the symphysis, and gradually rises in its middle portion into a high and stout coronoid process, which inclines a little backwards and ends abruptly behind. Beyond this process the tapering posterior end of the bone is much shorter than in the corresponding part of *Leptolepis*. The lower half of the bone is bent inwards along an obtuse-angled longitudinal ridge beginning below the middle of the symphysis, and this face is marked by large pits and a groove, which indicate a considerable development of slime-apparatus. The outer face is for the most part smooth, but there is a little rugosity near the oral border, which (as shown in Pl. XXIV, fig. 4 a) bears small, hollow, smooth, and bluntly-conical teeth.

The hyoid arch, seen in Pl. XXIV, fig. 4, is relatively large, the ceratohyal (*ch.*) being laterally compressed, mesially constricted, and deepest behind at its articulation with the epihyal (*eph.*). Its upper angles seem to be united by a rod of bone as in *Leptolepis*. The basihyal (*bh.*) is very short.

The preoperculum (Pl. XXV, fig. 1, *pop.*) is sharply angulated, with the lower limb nearly as long as the upper limb, and the anterior border much thickened. This thickening is smooth and widest at the bend, from which a few coarse ridges, more or less fused into a reticulation, radiate backwards over the thin triangular expansion. The traversing slime-canal is very large. The operculum (*op.*) seems to have been two-thirds as wide as deep, and is also thickened along its anterior border, from which at the point of suspension a short ridge diverges at an acute angle downwards and backwards, as if to bound a slime-pit (Pl. XXV, fig. 1 a). The outer face is entirely smooth. The suboperculum (*sop.*) and interoperculum are comparatively small, and both these and the broad upper branchiostegal rays are smooth. The lower branchiostegal rays, of which some are seen below the ceratohyal in Pl. XXIV, fig. 4 (*br.*), are slender, rod-shaped, and spaced.

As shown in several specimens, but especially well in the original of Pl. XXIV, fig. 4, the gill-arches bear a close series of large bony gill-rakers (*g.r.*) which are smooth, laterally-compressed elongated cones, each with a notch just above its base of attachment. Similar gill-rakers appear to occur in *Leptolepis* (B.M. no. P. 3674), *Æthalion* (B.M. no. 37042), and *Thrissops* (B.M. no. P. 917).

The vertebræ in the type specimen (Pl. XXV, fig. 1) are in undisturbed series, but they are somewhat broken, and not so well seen as in some of the portions of larger individuals. About 35 can be counted in the abdominal, 26 in the caudal region. The centra are all well ossified, the primitive double cone being thick and forming two wide rims, between which the secondary ossification is in fine longitudinal ridges, producing a striated appearance on superficial view. When a centrum is isolated and exposed on the articular face (Pl. XXV, fig. 3), a median perforation is seen for a persistent strand of the notochord. All the abdominal vertebral centra are deeper than long, with gently rounded sides, and their appended arches are loosely articulated. The foremost centrum is not united with the basioccipital, but (as shown in Pl. XXIV, fig. 4) it is peculiar in consisting of two fused discs, which are limited by a sharp line, the front disc bearing a slight prominence above for the support of the first neural arch, the second bearing a similar pair of slight prominences below to carry the first pair of ribs. The second centrum (sometimes also the third centrum) again exhibits prominences, but all the other abdominal centra are pitted for the reception of both the neural arches and ribs. The caudal vertebral centra are more elongated, about as long as deep, and their secondary ossification is disposed so as to form a sharp median longitudinal ridge on each side. Their appended arches are more firmly fixed in the sockets than those in the abdominal region, even if they are not fused. The arches are almost destroyed in the type specimen, but they are seen in others, as in the fragment represented in Pl XXIV, fig. 5 The neural arches are comparatively small and delicate, and unite by large anterior zygapophyses, but they are obscured in the abdominal region by the numerous well-developed intermuscular bones which overlie them. In the caudal region the neural and hæmal arches are nearly symmetrical, and sharply curved backwards, and a few intermuscular bones occur both above and below the vertebral centra. The stout gently-curved ribs extend to the ventral border of the fish, each with a slightly expanded articular head and a wide groove along its anterior or outer face.

The stout clavicle, of which the upper half is shown in Pl. XXV, fig. 1 (*cl.*), bears a large lateral expansion which is nearly smooth, only marked by a few vertical ridges and grooves or wrinkles at the upper end. The long and narrow supraclavicle (*scl.*) is similarly ridged, and traversed by a large slime-canal. The pectoral fin must have been relatively large, its rays when adpressed to the trunk (as in the type specimen) reaching two-thirds of the distance to the insertion of the pelvic fins. The pelvic fin-supports (Pl. XXIV; fig. 5, *plv.*) are long and narrow laminæ, widest at the articular end, tapering forwards, strengthened by a rib-like thickening along the outer edge, and apparently fixed together along the thin inner edge. The pelvic fin-rays, about 10 in number, are deeply imbricating and closely articulated distally. The dorsal and anal fins are shown in the type specimen to be acuminate in front and directly opposed, but their details are better seen in other specimens. In the dorsal fin there are about 18 rays, of which the anterior four are simple and gradually increase in length, while the fifth is the longest, and this and the following are divided and articulated distally (B.M. no. P. 4536). The foremost support is fan-shaped, bearing the first three rays (Pl. XXIV, fig. 5, *d.*), and the other supports are winged at the articular end.

The anal fin is as deep as the dorsal and closely similar, but more extended, probably with not less than 23 rays. The long foremost support is rod-shaped, and the wings at the articular ends of the other supports are short. The forked caudal fin is well shown in the type specimen, with the step-shaped articulations of its thicker rays. All the hæmal spines within the base of the caudal fin are stout, and two bearing the middle rays are slightly expanded. There are no traces of fulcra on any of the fins.

Remains of rather large cycloidal scales are seen in the original of Pl. XXIV, fig. 5; but they must have been very thin, and are usually absent in the fossils.

Remarks.—When examples of *Pachythrissops lævis* were first described, they were provisionally ascribed to the Wealden species treated below; but the form is readily distinguished by the characters of the opercular apparatus and the sharpness of the lateral ridge on the caudal vertebræ, besides by the proportions of the head.

Horizon and Locality.—Middle Purbeck Beds: Swanage, Dorset.

2. **Pachythrissops vectensis**, A. S. Woodward. Plate XXIV, figs. 6, 7; Plate XXV, figs. 4, 5; Text-figure 40.

1890. *Oligopleurus vectensis*, A. S. Woodward, Proc. Zool. Soc., p. 346, pl. xxviii, fig. 1.
1911. *Oligopleurus vectensis*, R. H. Traquair, Poiss. Wealdiens de Bernissart (Mém. Mus. Roy. Hist. Nat. Belg., vol. v), p. 47, pl. x.
1913. *Oligopleurus vectensis*, E. S. Goodrich, Proc. Zool. Soc., p. 84.

Type.—Head; British Museum.

Specific Characters.—Attaining a length of a little more than a metre. Length of head with opercular apparatus about equal to the distance between the paired fins. Operculum apparently without slime-pit, smooth, or faintly rugose, but fimbriated at the postero-superior border; preoperculum with a few strong radiating ridges spaced over the lower limb. Vertebræ 35 in abdominal region; centra in caudal region not longer than those in abdominal region, without any sharp lateral ridge, but slightly indented above and below to produce a broad rounded lateral ridge. [Fins imperfectly known.] Scales ornamented with sparse pustulations, and more or less fimbriated at the hinder border.

Description of Specimens.—The type specimen is an imperfectly preserved skull and mandible shown of one-half nat. size from the right side in Text-fig. 40. The greater part of a fish (Pl. XXV, fig. 4) discovered by Mr. Reginald W. Hooley, F.G.S., exhibits the head and most of the vertebral axis, besides some remains of fins. Other more fragmentary specimens make known a few additional details.

The skull is always more or less crushed and broken, but many of its principal features are shown in the fossils. As seen in the type specimen, there is a well-ossified supraoccipital, bearing a vertical median crest on its posterior face; and on either side of this there are remains of an equally well-ossified epiotic. These bones are exposed as a narrow rim at the posterior border of the cranial roof. As in *P. lævis*, each parietal bone is longer than wide, and the squamosal is relatively wide, though not much longer than the parietal. The large frontal bones, as shown in the type specimen and in Mr. Hooley's fish (Pl. XXV, fig. 4), are narrowed between the orbits; but, as also shown in Mr. Hooley's specimens, they are much expanded in the postorbital region and overlap the well-ossified postfrontals to an undetermined extent. The parietals, squamosals and frontals are marked only by fine radiating ridges and an occasional trace of the slime-canal,

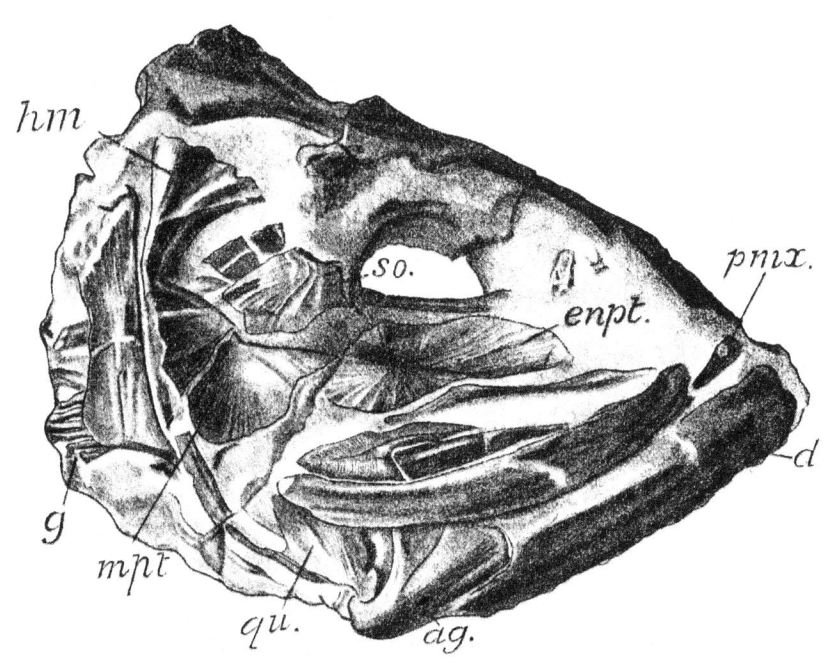

Fig. 40.—*Pachythrissops vectensis*, A. S. Woodward; type skull, right lateral view, one-half nat. size.—Wealden; Isle of Wight. British Museum, no 42013. *ag.*, articulo-angular; *d.*, dentary; *enpt.*, entopterygoid; *g.*, gill-rakers; *hm.*, hyomandibular; *mpt.*, metapterygoid; *pmx.*, premaxilla; *qu.*, quadrate; *so.*, portion of hinder cheek-plate.

but along the middle of the cranial roof there is a well-marked depression, which is widest and deepest between the hinder part of the orbits. The type specimen proves that the ethmoid region is comparatively small. Fragments of very thin cheek-plates occur in the type specimen (Text-fig. 40, *so.*) and in the original of Pl. XXV, fig. 4; and a single broad series is seen to cover the postorbital region of the cheek in Pl. XXIV, fig. 6, *so.* These plates are nearly smooth, but are marked with slight radiating ridges or fimbriations. They are also traversed by the slime-canal near the orbital border. The mandibular suspensorium curves forwards so that the articulation for the mandible is beneath the hinder part of the orbit. The hyomandibular (Pl. XXV, fig. 5) is much expanded at the upper end, with an apparently double-headed articulation; and, as in other fishes in which the preoperculum is relatively large, the process (*p.*) for the suspension of the operculum is much elongated. Just below the latter process the bone is antero-posteriorly compressed and rises to a sharp vertical crest along its outer face. The triangular quadrate (Text-fig. 40, *qu.*) is cleft postero-superiorly for the rod-shaped symplectic (Pl. XXV, fig. 4, *sy.*), and articulates above with a large triangular metapterygoid (*mpt.*). Its articular condyle for the mandible is somewhat constricted from the main part of the bone, and has a robust inwardly-directed process arising from its base. The entopterygoid (*enpt.*) is an elongated thin lamina of bone, bluntly pointed in front, truncated behind, slightly convex on the oral face, and inclined inwards towards the parasphenoid. The ectopterygoid (Pl. XXV, fig. 4, *ecpt.*) is a narrower elongated lamina, more rapidly tapering in front. It is toothless along the lower border, but its oral face, like that of the other pterygoid plates, has not been seen. The premaxilla, imperfectly shown in the type specimen (Text-fig. 40, *pmx.*), is comparatively small, but longer than deep. The large maxilla (Text-fig. 40 and Pl. XXV, fig. 4, *mx.*) is arched, and its outer face is nearly smooth, marked only in places with a faint rugosity; its convex oral border exhibits the points of attachment of clustered minute teeth. Its posterior two-thirds are overlapped by two supramaxillæ, of which the posterior is the larger and excavated in front for the anterior, which is very long and narrow and tapers forwards. The outer face of the posterior supramaxilla is slightly rugose, and bears a ridge along the lower margin of its antero-superior extension. The articulo-angular bone of the mandible (Text-fig. 40, *ag.*) is short and deep, and its lower portion is ornamented with fine longitudinal rugæ. The very large dentary (*d.*), which is imperfectly known, is evidently truncated at the symphysis, and its lower half is sharply bent inwards, separated from the upper half by an obtuse longitudinal ridge.

The hyoid arch, partly seen in the original of Pl. XXV, fig. 4, is relatively large, and the upper branchiostegal rays, attached to the epihyal, are especially large smooth laminæ of bone (*br.*).

The preoperculum, incomplete and abraded in the type specimen, is best seen in Pl. XXIV, figs. 6 (*pop.*), 6 *a*. It is sharply angulated, with a relatively large lower limb, and a great smooth expansion behind. The anterior border is much thickened and smooth, and from this thickening six or seven coarse ridges radiate over the outer face of the lower limb. The operculum (*op.*) is imperfect in all the specimens, and the proportions of the suboperculum and interoperculum are uncertain. The operculum seems to have been smooth or only faintly rugose, except at its postero-superior border, where it is finely fimbriated.

A single vertebral centrum attached to the occipital region of the type specimen is completely ossified, deeper than wide, and very short, marked on the sides by fine striations which extend between a thickened rim anteriorly and posteriorly. The neural and hæmal arches are inserted in pits. As shown by the large fish (Pl. XXV, fig. 4), all the vertebral centra are of this type, except that they are generally longer in proportion to their depth. There seem to be 34 or 35 in the abdominal region, but the number of caudals is uncertain. The centra in the caudal region are scarcely longer than those in the abdominal region, but they differ in exhibiting a slight lateral indentation above and below a rounded median lateral ridge. The neural arches in the abdominal region are long and slender, but most of them are broken and displaced, and obscured by the overlapping intermuscular bones. The stout, gently-arched ribs are well shown, reaching the ventral border, each impressed by a longitudinal groove. The neural and hæmal arches in the caudal region are firmly fixed in their sockets in the centra.

The remains of the clavicle (Pl. XXV, fig. 4, *cl.*) and supraclavicle (*scl.*) include a wide, nearly smooth, exposed portion; and a large smooth plate of bone (*x.*) shown behind the skull is probably the post-temporal, much expanded, as in *Thrissops*. The long pectoral fin-rays (*pct.*) are very closely articulated distally. The pelvic fins (*plv.*), with slender supports, are much smaller than the pectorals, and are evidently inserted nearer to the anal than to the latter. Of the anal fin (*a.*), only a fragment remains with at least 17 rays. As the anterior supports are relatively long and stout, this fin must have been acuminate. The dorsal and caudal fins are unknown.

Scales are seen in some of Mr. Hooley's specimens, all large, thin, and cycloidal, and deeply overlapping. They are most conspicuously marked by the concentric lines of growth; but some exhibit sparse pustulations, usually in rows, on the smooth exposed face, besides fine fimbriations at the hinder margin (Pl. XXIV, fig. 7).

Horizon and Localities.—Weald Clay: Atherfield, Isle of Wight; Berwick, Sussex.

Genus THRISSOPS, Agassiz.

Thrissops, L. Agassiz, Poiss. Foss., vol. ii, pt. i, 1833, p. 12.

Generic Characters.—Head as in *Leptolepis*, but smaller, and the dentary bone with a broader, less thickened elevation. Vertebral centra much strengthened by secondary ossification in longitudinal ridges, of which the median lateral is usually very prominent; no centra much deeper than long. Ribs especially stout. Pelvic fins relatively small; dorsal and anal fins acuminate and opposite, the former small and short-based, the latter much extended; caudal fin forked. Scales thin, completely covering the trunk; no enlarged or thickened ridge-scales.

Type Species.—*Thrissops formosus* (L. Agassiz, Poiss. Foss., vol. ii, pt. i, 1833,

p. 12; vol. ii, pt. ii, 1844, p. 124, pl. lxv a), from the Lithographic Stone (Lower Kimmeridgian) of Bavaria.

Remarks—*Thrissops* seems to range in Europe from the Middle Jurassic (Oxfordian) to the Lower Cretaceous; but without a knowledge of their cranial osteology, the systematic position of the later species usually ascribed to this genus must remain uncertain. They closely approach the Ichthyodectidæ or Chirocentridæ.

1. **Thrissops curtus,** sp. nov. Plate XXVI, fig. 1.

Type.—Imperfect fish; British Museum

Specific Characters.—At least 16 cm. in length. Length of head with opercular apparatus nearly equalling maximum depth of trunk, and contained from four to five times in total length of fish; maximum depth of trunk contained about two-and-a-half times in length from pectoral arch to base of caudal fin. Pelvic fins half as large as pectorals, inserted nearer to the anal fin than to the latter; dorsal fin, with 13 rays, arising behind the origin of the anal fin, which is slightly more elevated in front and comprises 30 rays; caudal fin very deeply forked and lobes slender.

Description of Specimen.—The type and only known specimen (Pl. XXVI, fig. 1) lacks the jaws, but is otherwise nearly complete. In the skull the ethmoidal region is very short and small, and the parasphenoid is seen crossing the relatively large orbit. A short ridge inclined upwards at the back of the occiput may perhaps be the thickened upper border of a triangular supraoccipital crest such as seems to occur in *Thrissops formosus* (Brit. Mus. no. P. 917). The preoperculum is vaguely shown to be much expanded at the angle and marked with several fine radiating ridges. The operculum, which is smooth, is widest below, and its maximum width is at least three-quarters of its depth. It seems to show a peculiar slime-pit near its point of suspension like that already described in *Pachythrissops lævis* (p. 131, Pl. XXV, fig. 1a). The subopercu1um, which is also smooth, is slightly more than one-third as deep as the operculum. The vertebral centra are evidently well ossified, but are not clearly seen except at the end of the tail, where they bear the usual sharp median lateral ridge. Over twenty vertebræ can be counted in the caudal region. The neural arches in the abdominal region are separated from the neural spines, which reach the dorsal border. Some of the anterior spines of this series have a small laminar expansion at their lower end. The stout grooved ribs, of which there are over twenty pairs, reach the ventral border. The neural and hæmal arches in the caudal region are fused both with their spines and with the corresponding centra; they are slender, short, and much inclined backwards. At least nine hæmals are included in the base of the caudal fin. Intermuscular bones occur only in the dorsal part of the abdominal region. Behind the skull, the post-temporal is a large, smooth, rhomboid plate, nearly as wide as deep, strengthened by a ridge extending diagonally from its antero-superior angle. The supraclavicle is relatively deep and narrow, and the clavicle is much arched. The anal fin comprises 27 rays, with three gradually lengthening fulcral rays in front, and its anterior acumination is about three-quarters as deep as the trunk at its insertion. The dorsal fin is slightly less elevated, with only 10 ordinary rays preceded by three fulcrals. Traces of the usual thin scales are seen, none with ornament.

Remarks.—Although the abdominal part of the vertebral column in the type specimen is somewhat displaced by crushing, the arrangement of the neural arches and ribs suggests that it has not been much shortened. The species is therefore peculiar in the shortness of the trunk, and it differs from the only known species from Portland (*T. portlandicus*, A. S. Woodward, Catal. Foss. Fishes, Brit. Mus., pt. iii, 1895, p. 525, pl. xviii, fig. 4) not only in this feature, but also in the more remote insertion of the dorsal fin.

Horizon and Locality.—Lower Purbeck Beds: Isle of Portland. The stratigraphical position was determined by the collector of the type specimen, Mr. J. R. Short. The exact horizon of *T. portlandicus* was not recorded by the collector.

2. **Thrissops molossus**, sp. nov. Plate XXVI, fig. 2.

Type.—Imperfect fish; British Museum.

Specific Characters.—At least 35 cm. in length. Head remarkably short and deep. Length of head with opercular apparatus nearly equalling maximum depth of trunk, and about one-fifth of total length of fish; maximum depth of trunk contained about three-and-a-half times in length from pectoral arch to base of caudal fin. Caudal fin very deeply forked and lobes slender.

Description of Specimen.—There is no doubt that the fragmentary fish shown of one-half the natural size in Pl. XXVI, fig. 2, represents a hitherto unrecognised species, but it is too imperfect for precise determination. The cranium is lacking, but the adjacent remains are sufficiently undisturbed to show that the head is unusually short and deep. The ossified border of the sclerotic of the very large eye is well preserved, and the mandibular suspensorium and jaws are also seen. The stout rod-shaped symplectic occurs below the hyomandibular, evidently fitting into a cleft of the large fan-shaped quadrate. The pterygoid bones form an extensive laminar expansion above the maxilla, which is gently arched and bears a regular close series of small conical teeth. Of the premaxilla only a fragment remains. The mandible is very imperfect, but its articulation is clearly below the hinder border of the orbit, and its oral border bears conical teeth which are larger than those of the maxilla. The remains of the operculum are bordered behind by part of the clavicle, and its width is thus shown to equal half the length of the head. The lower branchiostegal rays are very slender. The parts of the trunk preserved seem to be almost in their natural position, but some of the abdominal vertebræ are displaced above the fossil, while the upper lobe of the caudal fin is torn away and displaced below. The axial skeleton is typically that of *Thrissops*, and the middle hæmal spine at the base of the caudal fin is much expanded. The stoutness and length of the anterior supports of the anal fin show that this must have been very deep and acuminate in front.

Horizon and Locality.—Middle Purbeck Beds: Swanage, Dorsetshire.

SUPPLEMENT.

Hybodus basanus, Egerton (p. 5). Plate XXVI, fig. 3.

Mr. Reginald W. Hooley, F.G.S., has obtained from the top of the Weald Clay in the typical locality, Atherfield, Isle of Wight, the well-preserved small head partly shown in Pl. XXVI, fig. 3. In the front part of the jaws it displays some of the characteristic teeth; and at the limits of the gape of the mouth there are remains of the labial cartilages, as already described (p. 7, Pl. II, figs. 1 *a*, 1 *b*). Even allowing for some vertical crushing, however, the skull clearly differs from the well-preserved specimen from Pevensey (Pl. II, fig. 1) in being relatively shorter and broader, with less widened postorbital processes. As seen from above (Pl. XXVI, fig. 3), the supraorbital flanges are well developed, and the interorbital width exceeds half the total length of the cranial roof. The posterior depression (*p.f.*) is less elongated than in the best-preserved Pevensey specimen. The small rostral prominence is evidently broken away by accident. The occiput (Pl. XXVI, fig. 3 *a*) is especially well shown, only abraded at its median vertical crest, which is still best seen in the skull already mentioned on p. 6. The occipital face is more than half as deep as wide, and its most conspicuous feature is the large excavation for the notochord (*n.*) in the basioccipital region, resembling that in *Notidanus* (*Hexanchus*).[1] The foramen magnum (*f.m.*), seen above this excavation, is comparatively small. A pair of deep pits (*v.*) in the cartilage flanking the foramen magnum would probably be pierced by the foramina of the vagus nerves.

Hybodus sulcatus, Agassiz.

The two fragments of a dorsal fin-spine named *Hybodus sulcatus* by Agassiz (Poiss. Foss., vol. iii, 1837, p. 44, pl. x *b*, figs. 15, 16), though originally stated by

[1] C. Gegenbaur, Untersuchungen zur vergleichenden Anatomie der Wirbelthiere.—III. Das Kopfskelet der Selachier (1872), p. 120, pl. iv, fig. 2, pl. xv, fig. 2.

Mantell to have been obtained from the Chalk of Lewes, are generally regarded as Wealden fossils (S. J. Mackie, The Geologist, vol. vi, 1863, p. 242, and A. S. Woodward, Catal. Foss. Fishes, Brit. Mus., pt. i, 1889, p. 275). A renewed examination of these fragments, however, suggests that they are from a ferruginous concretion in the Chalk, and not from any Wealden deposit. As already suspected by M. Leriche (Mém. Soc. Géol. Nord, vol. v, 1906, p. 91), they are parts of the dorsal fin-spine of a Chimæroid fish; but they still remain unique. The anterior border of the spine is less compressed to an edge than usual, and it is ornamented with a few large and irregularly arranged tubercles (not shown in Agassiz' figures). The longitudinal ribbing of the lateral faces is coarser and more marked than in any other known Chimæroid spine. The specimen may perhaps belong to a species of *Ischyodus*.

Hybodus subcarinatus, Agassiz (p. 10).

Lepidotus mantelli, Agassiz (p. 36).

According to Mr. J. Wilfrid Jackson, the type specimens of *Hybodus subcarinatus* and the so-called *Tetragonolepis mastodonteus* (= small dentary of *Lepidotus*) are in the Cumberland Collection in the Manchester Museum.

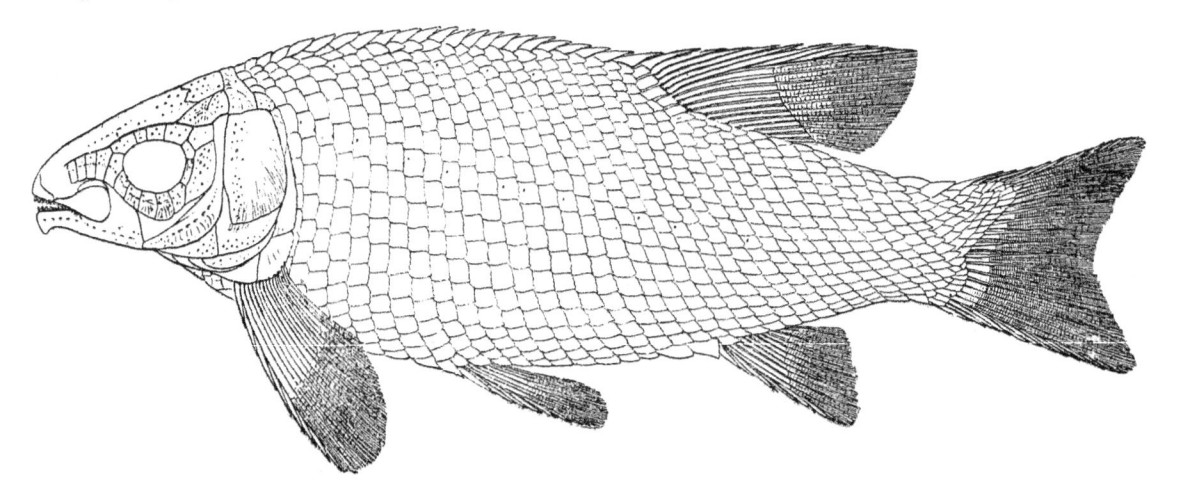

Fig. 41.—*Lepidotus minor*, Agassiz; restoration with amended dorsal and anal fins, to replace the restoration given in Text-fig. 14, p. 28.

Lepidotus minor, Agassiz (p. 27). Plate XXVI, fig. 4.

Dr. F. Du Cane Godman, F.R.S., has lately given to the British Museum the large specimen of the stout variety of *Lepidotus minor*, from the Middle Purbeck Beds of Swanage, shown of one-third the natural size in Pl. XXVI, fig. 4. It is important as displaying the complete dorsal fin, which is proved to have been wrongly drawn in the restoration of the species in Text-fig. 14, p. 28. A new restoration with the dorsal and anal fins amended is accordingly given in Text-fig. 41. The specimen is much laterally compressed by crushing, so that the left half of the cranial roof is bent downwards and has slipped partly beneath the right half. The mouth is also opened, and the ventral part of the abdominal region is displaced downwards. The large characteristic smooth maxilla (*mx.*) is well seen, with part of the supramaxilla, and the ectopterygoid (*ecpt.*) and quadrate (*qu.*) are exposed. The small upper postorbital cheek-plate seems to be subdivided by a vertical suture; the large postorbital is also divided transversely in its lower half, but this may be merely an accidental crack. The tubercular ornament of the opercular bones is unusually extensive and dense. Remains of the paired fins show their relative proportions. The enamelled fulcra in the dorsal fin are slightly curved, forming a convexly arched border; the closely divided and articulated fin-rays seem to be complete, and resemble those of *L. mantelli* (see Pl. VII, fig. 7). Some of the anterior scales of the flank are coarsely serrated in their lower half, but further back the corresponding scales do not exhibit more than one denticulation above the produced postero-inferior angle, and this soon disappears at the beginning of the tail. From the produced angle, and sometimes from the denticulation above, a faint oblique ridge extends forwards over the scale. The upper slime-canal, which begins as usual on the third row of scales below the dorsal ridge, descends and becomes a little irregular near the dorsal fin.

SUMMARY AND CONCLUSION.

Little is added to our knowledge of the Wealden and Purbeck fish-fauna by the fossils discovered in the corresponding formations on the European continent. Even in the great collection described by Traquair,[1] from the Wealden of Bernissart, Belgium, there is only one genus, the Macrosemiid *Notagogus*, additional to those represented among the English fossils. At Bernissart no Selachian remains are known, but in north Germany[2] there are teeth and spines of *Hybodus* much resembling those now described. In north Germany several fine specimens of *Lepidotus* have also been discovered,[3] but of other ganoids there are only teeth and fragmentary jaws, chiefly referable to Pycnodonts.

So far as known, therefore, the fishes of the Wealden and Purbeck formations are essentially Jurassic, and not mingled with any typically Cretaceous forms. Most of them are, indeed, the specialised and evidently final representatives of the Jurassic families to which they belong, and very few can be regarded as possible ancestors of fishes which followed in Cretaceous and later times.

[1] R. H. Traquair, " Les Poissons Wealdiens de Bernissart," Mém. Mus. Roy. Hist. Nat. Belgique, vol. vi (1911), pp. 1—65, pls. i—xii.

[2] W. Dunker, Monographie der Norddeutschen Wealdenbildung. Brunswick, 1846.

[3] W. Branco, " Beiträge zur Kenntniss der Gattung *Lepidotus*," Abhandl. geol. Specialk. Preussen u. Thüring. Staaten, vol. vii (1887), pt. 4, pp. 1—84, pls. i—viii.

Among the Selachians, the well-preserved skulls of *Hybodus basanus* are especially interesting, because they exhibit much more resemblance to those of the Notidanidæ than to the skull of the modern *Cestracion*, to which *Hybodus* is commonly regarded as nearly allied. The teeth of some of the Lower Jurassic species of *Hybodus*, indeed, seem to pass into those of the earliest known species of *Notidanus*[1]; just as the high-cusped teeth of the Wealden *H. basanus* are closely similar to some of the Middle and Upper Jurassic teeth of *Orthacodus*,[2] which seem to pass into those of the primitive though typical Lamnidæ of the Cretaceous period. Some of the more generalised Hybodonts, when better known, may therefore prove to be ancestral to several later types of sharks.

Hybodus basanus is also noteworthy in having only one pair of hooked cephalic spines instead of the two pairs, of nearly equal size, which characterise the early Liassic species.[3] In one species of the allied genus *Asteracanthus*, from the Oxford Clay of Peterborough, the late Mr. Alfred N. Leeds observed that of the two pairs of cephalic spines one was much smaller than the other. There was thus probably a similar reduction of one pair in some species of *Hybodus*, which eventually resulted in its complete loss.

The occurrence of a dwarf species of *Acrodus* in the Wealden is paralleled by that of a similar dwarf species in an estuarine deposit of nearly the same age in Bahia, Brazil.[4] This is almost the last appearance of the genus, the latest known species being *Acrodus levis* from the Gault.

No remains of Crossopterygians have hitherto been found in the Wealden, and the only specimen from the Purbeck Beds is that of the typically Jurassic *Undina* now described. The ornament of the scales clearly distinguishes it from the Cretaceous *Macropoma*.

Among Chondrosteans the Palæoniscidæ are represented for the last time by the genus *Coccolepis*, which is too specialised to be ancestral to the later sturgeons.

The representatives of *Lepidotus* are interesting as including a comparatively generalised species, *L. minor*, which might even be Lower Jurassic, besides a large and highly specialised species, *L. mantelli*, which could not be earlier than Upper Jurassic, and might be Lower Cretaceous. The vertebral ossifications in the latter species are particularly remarkable.

The Pycnodonts also comprise both generalised and moderately specialised

---

[1] Compare *Hybodus polyprion*, Ag., with *Notidanus muensteri*, Ag. (A. S. Woodward, Geol. Mag., [3] vol. iii (1886), p. 257, pl. vi, figs. 1—5).

[2] O. Jaekel, Sitzungsb. Gesell. naturf. Freunde, Berlin, 1898, p. 139, text-fig. 2; F. Priem, Bull. Soc. Géol. France, [4] vol. xii (1912), p. 254, with figs. A large tooth apparently of *Orthacodus*, from the Danian of Scandinavia, has been described as *Oxyrhina lundgreni*, J. W. Davis, Trans. Roy. Dublin Soc., [2] vol. iv (1890), p. 393, pl. xxxix, figs. 8—13.

[3] See *Hybodus delabechei* (Charlesworth), A. S. Woodward, Catal. Foss. Fishes, Brit. Mus., pt. i (1889), p. 259, pl. viii, fig. 1.

[4] *Acrodus nitidus*, A. S. Woodward, op. cit. (1889), p. 297, pl. xiv, fig. 8.

forms, but no exclusively Cretaceous genera. The specimens of *Microdon radiatus* and *Mesodon parvus* from the Purbeck Beds are especially well preserved to show the osteology, and a study of them has led to several new observations.[1] There can no longer be any doubt as to the close relationship of these fishes to the *Lepidotus*-like ganoids.

The Purbeckian specimens of *Ophiopsis* and *Histionotus* make some additions to our knowledge of the osteology of the more generalised Macrosemiidæ, which emphasise their affinity to the Eugnathidæ. The curious development of large slime-pits in the preoperculum and supratemporal bones of *Histionotus* still awaits explanation. If the new genus *Enchelyolepis* be rightly placed in the same family, its very thin cycloidal scales are particularly interesting and novel.

Besides the remains of typical Eugnathidæ of Jurassic facies, one example of the later Cretaceous genus *Neorhombolepis* is described from the Wealden. Either this, however, or a nearly similar genus also occurs in the apparently Lower Cretaceous estuarine deposit in Bahia, Brazil,[2] already mentioned, while the closely-related *Otomitla* is found in the Neocomian of Mexico.[3]

Amiopsis, though so closely similar to *Amia*, still retains the short dorsal fin which distinguishes the Jurassic members of the family. It is a Cretaceous genus.

The Purbeckian *Aspidorhynchus fisheri* seems to be the latest species of the genus hitherto discovered. The specimens now described show much of the osteology, and suggest that the supposed close relationships of the Aspidorhynchidæ to the Lepidosteidæ need further examination. The Wealden remains of *Belonostomus* are unfortunately too fragmentary for discussion.

The Pholidophoridæ occur for the last time, and the Purbeckian species of *Pholidophorus* are interesting for the strength of their fins. *P. ornatus* and *P. granulatus* are especially similar to species from the Upper Jurassic Lithographic Stone of Germany and France. The almost scaleless *Ceramurus* may be regarded as a highly specialised member of the family. Some of the Purbeckian specimens of *Pleuropholis* are in an unusually good state of preservation, and add to our knowledge of the osteology of the genus.

Among the Leptolepidæ, which are also essentially Jurassic fishes, may perhaps be recognised some of the ancestors of the typical physostomous teleosteans of the later Cretaceous seas. They evidently connect the more typical "ganoids" with the primitive Elopidæ which are so abundant in marine Cretaceous formations. By a relative enlargement of the supraoccipital and otic bones in the skull, a

[1] A. S. Woodward, "Notes on the Pycnodont Fishes," Geol. Mag., [6] vol. iv (1917), pp. 385—389, pl. xxiv.

[2] A. S. Woodward, "The Fossil Fishes of the English Chalk" (Palæont. Soc., 1911), p. 256.

[3] *Otomitla speciosa*, J. Felix, Palæontographica, vol. xxxvii (1891), p. 189, pl. xxix, fig. 3, pl. xxx, figs. 3—5.

partial fusion of the hæmal spines at the base of the caudal fin, and an increased development of the intermuscular bones, some of them may have become Clupeidæ. By nearly the same changes, such genera as *Pachythrissops* and *Thrissops* may also have passed into the *Chirocentrus*-like fishes, which were as numerous in the Cretaceous fauna as the Elopines and Clupeoids, and are represented at the present day by a genus (*Chirocentrus*) which retains a remnant of the primitive spiral valve in the intestine.[1]

It is therefore interesting to compare the fishes of the Wealden and Purbeck estuary with those of the contemporaneous seas in the European area. It is only unfortunate that the latter are thus far very imperfectly known. Fish-remains are not uncommon in some of the marine Neocomian formations of France which seem to be contemporary with at least the later Wealden deposits, but they are all very fragmentary.[2] So far as determinable, nearly all of them are typically Jurassic; though it should be remembered that these are durable fossils such as the teeth and fin-spines of sharks, and the teeth and jaws of Pycnodonts and *Lepidotus*. More delicate skeletons may have been destroyed beyond recognition, as suggested by the discovery of an otolith which may be Clupeoid.[3] Among these fossils, however, both in France and in Switzerland there are a few Selachian teeth[4] so closely similar to those of the Lamnidæ that they evidently indicate the appearance of the modern type of shark which became so abundant and widely spread in the later Cretaceous seas. Well-preserved fishes in another Neocomian formation in the Voirons, Switzerland,[5] include recognisable forerunners of the Chirocentrids (*Spathodactylus neocomiensis*) and Clupeoids (*Crossognathus sabaudianus*, *Clupea antiqua*, and *Clupea voironensis*), those referred to *Clupea* itself showing distinctly the characteristic ventral ridge-scutes. The Clupeoid *Crossognathus* also occurs in the Hilsthon of Hanover, which seems to correspond with the uppermost part of the Weald Clay. The marine fish-fauna of Europe before the end of Wealden times was thus distinctly in advance of the estuarine fauna so far as known, and approached the later Cretaceous fauna in its Lamnid sharks, Chirocentrids, and Clupeoids.

The nearly contemporaneous estuarine deposit on the coast of Bahia, Brazil, to which reference has already been made, shows that the Lower Cretaceous fish-

---

[1] Cuvier and Valenciennes, Histoire Naturelle des Poissons, vol. xix (1846), p. 160, pl. 565.

[2] M. Leriche, " Sur quelques Poissons du Crétacé du Bassin de Paris," Bull. Soc. Géol. France, [4] vol. x (1910), pp. 455—469, pl. vi.

[3] *Otolithus* (*Clupeidarum ?*) *neocomiensis*, F. Priem, " Poissons Fossiles du Bassin Parisien " (Publ. Ann. Paléont., 1908), p. 37, text-figs. 11—14.

[4] *Lamna* (*Odontaspis*) *gracilis*, L. Agassiz, Poiss. Foss., vol. iii (1843), p. 295, pl. xxxviia, figs. 2—4. *Odontaspis macrorhiza* mut. *infracretacea*, M. Leriche, Bull. Soc. Géol. France, [4] vol. x. (1910), p. 459.

[5] F. J. Pictet, " Description des Fossiles du Terrain Néocomien des Voirons " (Matér. Paléont. Suisse, ser. 2, 1858), pp. 1—54, pls. i—vii.

fauna of that region was essentially the same as in Europe. According to the discoveries of Mr. Joseph Mawson,[1] it includes such typical survivors of the Jurassic fauna as *Acrodus*, *Lepidotus*, *Megalurus*, and probably *Belonostomus*, besides a few more advanced forms, among which a Chirocentrid (*Chiromystus mawsoni*) and a Clupeoid (*Diplomystus longicostatus*) are clearly recognisable. The only unique fish found here is a gigantic Cœlacanth, *Mawsonia*, as highly specialised as the Upper Cretaceous *Macropoma*. The fish-remains hitherto recorded from the supposed Neocomian of Mexico,[2] the United States,[3] and East Africa[4] are also of an essentially Jurassic facies. It is therefore evident that a specially rapid evolution of sharks, skates, and teleosteans occurred in Middle Cretaceous times.

In conclusion, the author desires to express his thanks to the many friends who have facilitated this work. He is especially indebted to the Director of the Geological Survey, Dr. F. L. Kitchin, and Mr. H. A. Allen; to the President and Committee of the Dorset Field Club, and Captain John E. Acland of the Dorset County Museum; to the late and present Woodwardian Professors at Cambridge, and Mr. Henry Woods of the Sedgwick Museum; to the Curators of the Museums of Devizes, Hastings, Manchester, Sherborne School, Warwick, and York; to the late Mr. Charles Dawson; and to Mr. Reginald W. Hooley. He also owes both the drawings on the plates and the new text-figures to Miss Gertrude M. Woodward.

---

[1] J. Mawson and A. S. Woodward, "On the Cretaceous Formation of Bahia (Brazil) and on Vertebrate Fossils collected therein," Quart. Journ. Geol. Soc., vol. lxiii (1907), pp. 128—139, pls. vi—viii. See also A. S. Woodward, Quart. Journ. Geol. Soc., vol. lxiv (1908), pp. 358—362, pls. xlii, xliii; and D. S. Jordan, Ann. Carnegie Mus. Pittsburgh, vol. vii (1910), pp. 23—34, pls. v—xiii.

[2] J. Felix, Palæontographica, vol. xxxvii (1891), pp. 189—194, pls. xxviii—xxx.

[3] E. D. Cope, Journ. Acad. Nat. Sci. Philad., [2] vol. ix (1894), pp. 441—447, pls. xix, xx; J. W. Gidley, Proc. U.S. Nat. Mus., vol. xlvi (1913), pp. 445—448, text-figs. 1—4.

[4] E. Hennig, Sitzungsb. Gesell. naturf. Freunde, Berlin, 1913, pp. 309, 315.

INDEX.

Note.—The roman and arabic numerals following a semicolon refer to the plates and figures illustrating those species which are recognised and described.

	PAGE
Acanthurus	29
Acrodus	14
— anningiæ	15
— gaillardoti	14
— hirudo	68
— levis	142
— nitidus	142
— nobilis	14
— ornatus	14; II, 15–18
Actinopterygii	23
Æthalion	125
— knorri	125
— robustus	126
— valdensis	126
— vidali	126
Amiidæ	90
Amiopsis	90
— austeni	94
— damoni	91; XIX, 5, 6
— dolloi	90, 91
— lata	90
— prisca	90
Aspidorhynchidæ	95
Aspidorhynchus	95
— acutirostris	95
— fisheri	97; XX, 1–4
Asteracanthus	16
— cephalic spines	142
— granulosus	18
— ornatissimus	16
— semiverrucosus	17
— verrucosus	16
Athrodon	47, 68
— douvillei	48
— intermedius	48
Attakeopsis (?) austeni	94
Belonostomus	100
— hooleyi	100; XXI, 1–3
— sphyrænoides	100
Bibliography	2

	PAGE
Callopterus	85
— insignis	83
— latidens	83
Caturus	82
— furcatus	82
— latidens	83
— purbeckensis	85; XIX, 1, 2
— tenuidens'	86; XIX, 3, 4
Ceramurus	111
— macrocephalus	111; XXII, 7
Cestraciontidæ	3
Chirocentridæ	144
Chirocentrus	144
Chiromystus mawsoni	145
Clupea antiqua	144
— knorri	125
— voironensis	144
Clupeidæ	144
Coccolepis	23
— andrewsi	24; IV, 2, 3
— australis	24
— bucklandi	23
— liassica	23
— macropterus	23
Cœlacanthidæ	21
Cœlodus	64
— arcuatus	70; XIII, 5
— costæ	65
— hirudo	68; XV, 14–18
— lævidens	69; XV, 19, 20
— mantelli	66; XV, 6–11
— multidens	67; XV, 12, 13
— saturnus	65
Conodus	82
Cosmodus	64
Crossognathus sabaudianus	144
Crossopterygii	21
Curtodus	16
Cyprinus elvensis	26
Diplomystus longicostatus	145
Ditaxiodus	82

INDEX.

	PAGE
Elasmobranchii	3
Elopidæ	143
Enchelyolepis	80
— andrewsi	80; XVII, 6
— pectoralis	81; XVII, 7
Endactis	82
Eomesodon	54
— barnesi	56; XIII, 1
— depressus	57; XIII, 2
— gibbosus	54
— granulatus	54
— hœferi	54
— liassicus	54, 55
— rugulosus	54
Esox acutirostris	95
Eugnathidæ	82
Eurystethus brongniarti	129
Glossodus	64
— mantelli	66
Gyrodus macropterus	49
— mantelli	66
Histionotus	76
— angularis	77; XVII, 1-5
Holophagus	21
Hybodus	3
— africanus	4
— basanus	5, 139; I, 1, 2; II, 1; XXVI, 3
— delabechei	142
— dorsalis	11
— ensis	11; II, 2-7
— fraasi	4
— grossiconus	11
— hauffianus	4
— obtusus	12
— parvidens	12; II, 8-14
— plicatilis	4
— polyprion	142
— reticulatus	4
— striatulus	12; III, 8
— strictus	13; III, 4, 5
— subcarinatus	10, 140
— sulcatus	139
— woodwardi	4
Hylæobatis	19
— problematica	19; V, 1-5

	PAGE
Lamna gracilis	144
Lepidotus	26
— elvensis	26
— fittoni	36, 37
— gigas	26
— lævis	47
— latifrons	29
— mantelli	27, 36; VII, 7; VIII, IX, X; XI, 1-14
— minor	27, 129, 140; V, 6-11; VI; VII, 1-5; XXVI, 4
— notopterus	34; VII, 6
— semiserratus	26, 27
— subdenticulatus	36
Leptolepidæ	121
Leptolepis	121
— brodiei	122; XXIII, 1-6
— bronni	122
— dubius	122
— nanus	125; XXIII, 2
— nathorsti	122
— neocomiensis	123
— valdensis	126
Macromesodon	48
Macrosemiidæ	70
Macrosemius	80
— andrewsi	80
— pectoralis	80
Mawsonia	145
Megalurus	90
— austeni	94
— damoni	91
— mawsoni	90
Megastoma	121
— apenninum	123
Meristodon	3
— paradoxus	10
Mesodon	48
— barnesi	56
— bernissartensis	49
— daviesi	50; XII, 1, 2
— gibbosus	54
— granulatus	54
— hœferi	54
— liassicus	49
— macropterus	49, 52
— parvus	52; XII, 3, 4
— rugulosus	54

INDEX.

	PAGE
Microdon	58
— elegans	59
— radiatus	59; XIV; XV, 1–5
Microps	101
Myliobatidæ (?)	19
Neorhombolepis	87
— excelsus	87
— valdensis	87; XVIII
Notidanus	142
— muensteri	142
Odontaspis gracilis	144
— macrorhiza mut. infracretacea	144
Œonoscopus	121
— cyprinoides	121
Oligopleuridæ	121
Oligopleurus	129
— vectensis	129, 133
Ophiopsis	70
— breviceps	73; XVI, 3–12
— dorsalis	75; XVI, 13
— penicillata	71; XVI, 1, 2
— procera	70, 71
Orthacodus	142
Otolithus (Clupeidarum?) neocomiensis	144
Otomitla	143
Oxygonius	121
— tenuis	125; XXIII, 1
Oxyrhina lundgreni	142
— paradoxa	10
Pachythrissops	128
— lævis	129; XXIV, 3–5; XXV, 1–3
— vectensis	133; XXIV, 6, 7; XXV, 4, 5
Palæoniscidæ	23
Parathrissops	128
— furcatus	129
Petalopteryx	80
Pholidophoridæ	101
Pholidophorus	101
— bechei	101
— brevis	110; XXII, 4, 5
— granulatus	106; XXI, 5, 6
— micronyx	103

	PAGE
Pholidophorus ornatus	102; XX, 5–8; XXI, 4
— purbeckensis	108; XXII, 1–3
Plesiodus	26
Pleuropholis	113
— attenuata	114
— crassicauda	118; XXIII, 12, 13
— formosa	115; XXII, 8; XXIII, 8–11
— longicauda	114, 119; XXIV, 1, 2
— serrata	114, 120
Prolepidotus	26
Ptychodus	19, 21
Pycnodontidæ	47
Pycnodus granulatus	54
— liassicus	54
— mantelli	66
— microdon	66
— rugulosus	54
Sarginites	121
— pygmæus	123
Scrobodus	26
Selachii	3
Semionotidæ	26
Spathodactylus neocomiensis	144
Sphærodus	26
Sphenonchus	3, 14, 16
— elongatus	14
Strobilodus	82
— purbeckensis	85
Strophodus	16
Teleostomi	21
Tetragonolepis mastodonteus	36, 42, 140
Tharsis	121
Thectodus	14
Thlattodus	82
Thrissops	136
— curtus	137; XXVI, 1
— formosus	136
— molossus	138; XXVI, 2
— portlandicus	138
Typodus	48
Undina	21
— penicillata	22
— purbeckensis	22; IV, 1
Uræus	82

ADLARD AND SON AND WEST NEWMAN, LTD., LONDON AND DORKING.

PLATE I.

FIG.   PAGE.

1. *Hybodus basanus*, Egerton; skull and mandible, right side view and (1 *a*) lower view, one-half nat. size, with a lower lateral tooth (1 *b*) enlarged twice.—Weald Clay; Atherfield, Isle of Wight. The type specimen. Museum of Practical Geology, London, no. 27973. *a.f.*, anterior fontanelle; *ch.*, ceratohyals; *l.*, labial cartilages; *md.*, mandible; *orb.*, orbit; *ptq.*, pterygo-quadrate (upper jaw).   6.

2. Ditto; partially decayed skull and mandible, left side view, one-half nat. size, with the cephalic spine (2 *a*) nat. size.—Weald Clay; Cooden Beach, Pevensey Bay, Sussex. Beckles Collection (B. M. no. P. 11871). *md.*, mandible; *ptq.*, pterygo-quadrate (upper jaw); *s.*, cephalic spine; *x.*, cartilage at base of cephalic spine.   9.

3. Hybodont Cephalic Spine; left side view, nat. size.—Wealden; Hastings. Rufford Collection (B. M. no. P. 6738).   14.

4. Hybodont Cephalic Spine; left side view and (4 *a*) hinder view, twice nat. size.—Wadhurst Clay; Brede, near Hastings. Teilhard & Pelletier Collection (B. M. no. P. 11895).   14.

PALÆONTOGRAPHICAL SOCIETY, 1915.

A. S. Woodward, Wealden & Purbeck Fishes.

Plate I.

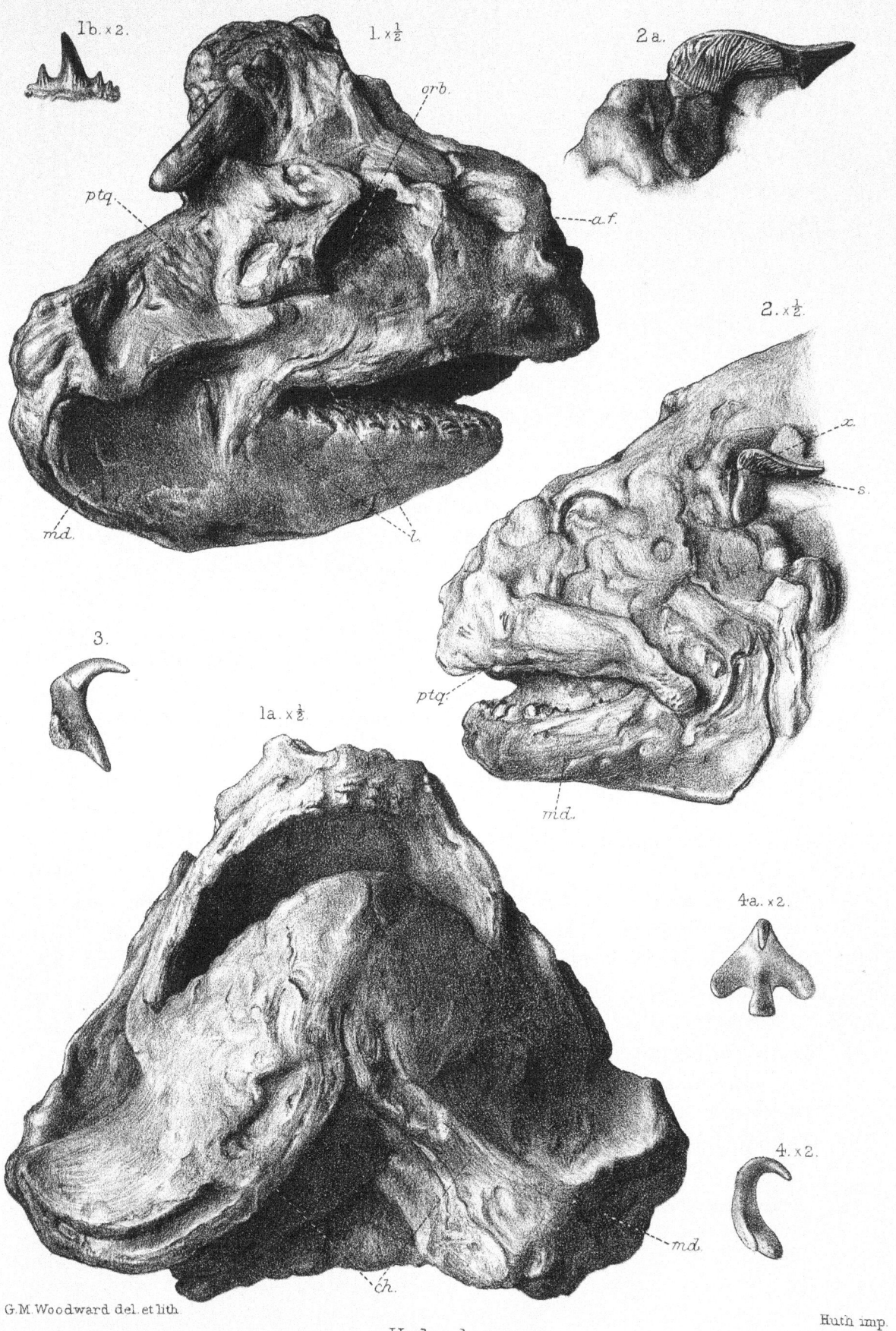

G. M. Woodward del. et lith.

Huth imp.

Hybodus.

PLATE II.

FIG.   PAGE.

1. *Hybodus basanus*, Egerton; skull and mandible, top view, left side view (1 *a*), and lower view (1 *b*), one-half nat. size.—Weald Clay; Cooden Beach, Pevensey Bay, Sussex. Beckles Collection (B. M. no. P. 11870). *a.f.*, anterior fontanelle; *ch.*, ceratohyal; *hm.*, hyomandibular; *l. l. 1, 2*, anterior and posterior lower labial cartilages; *md.*, mandible; *orb.*, orbit; *p.f.*, posterior fontanelle; *ptq.*, pterygoquadrate (upper jaw); *r.*, rostrum; *u. l. 1, 2*, anterior and posterior upper labial cartilages.   6.

1 *c, d*. Ditto; dermal tubercles in side view (1 *c*) and lower view (1 *d*), ten times nat. size.—Ibid. Beckles Collection (B. M. no. P. 11872).   10.

2. *Hybodus ensis*, sp. nov.; tooth with broken apex, nat. size.—Middle Purbeck Beds; Swanage, Dorset. Museum of Practical Geology, London, no. 27979.   11.

3. Ditto; crushed tooth, nat. size.—Ibid. M. P. G. no. 27976.   11.

4. Ditto; tooth, nat. size.—Middle Purbeck Beds, Durlston Bay, Swanage. York Museum.   11.

5. Ditto; imperfect large tooth, abraded, nat. size.—Middle Purbeck Beds; Swanage. M. P. G. no. 27978.   11.

6. Ditto; tooth, twice nat. size.—Ibid. The type specimen. B. M. no. 21349.   11.

7. Ditto; postero-lateral tooth, three times nat. size.—Ibid. B. M. no. 21349 *b*.   11.

8. *Hybodus parvidens*, sp. nov.; tooth, three times nat. size.—Wadhurst Clay; Hastings. The type specimen. Teilhard & Pelletier Collection (B. M. no. P. 11877).   12.

9–14. Ditto; six teeth, three times nat. size.—Ibid. Teilhard & Pelletier Collection (B. M. nos. P. 11878–83).   12.

15, 16. *Acrodus ornatus*, A. S. Woodward; two principal teeth, five times nat. size.—Wealden; Brook, Isle of Wight. Sedgwick Museum, Cambridge.   15.

17. Ditto; anterior tooth, four times nat. size.—Ibid. Sedgwick Museum, Cambridge.   15.

18. Ditto; posterior tooth, five times nat. size.—Wealden; Bexhill, Sussex. B. M. no. P. 6105.   15.

PALÆONTOGRAPHICAL SOCIETY, 1915.

A. S. Woodward, Wealden & Purbeck Fishes.

Plate II.

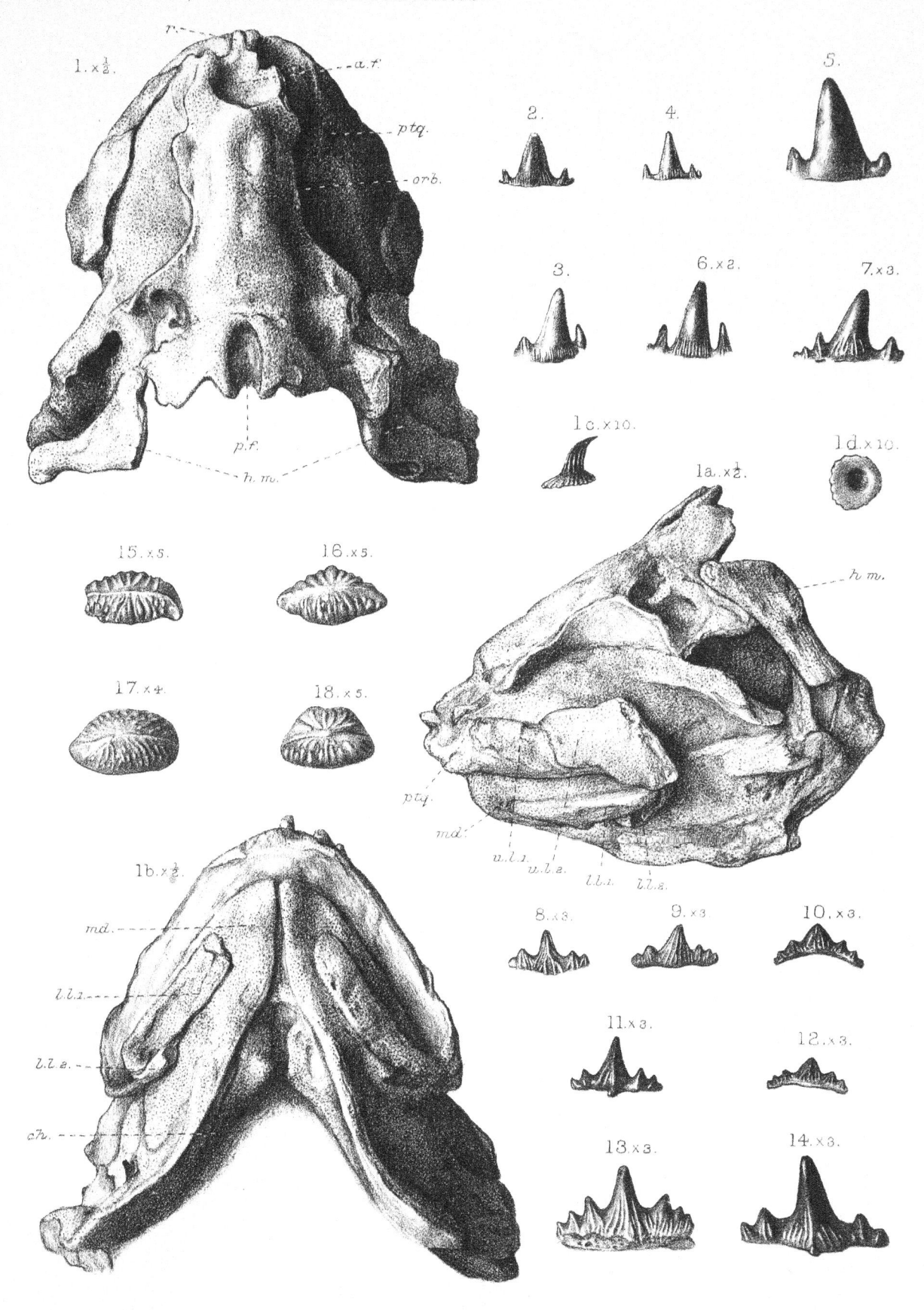

1–14 Hybodus.   15–18 Acrodus

PLATE III.

Fig.		Page.
1.	*Hybodus ensis*, sp. nov. (?); dorsal fin-spine.—Middle Purbeck Beds; Swanage, Dorset. B. M. no. 46908.	11.
2.	Ditto; imperfect large dorsal fin-spine.—Ibid. Dorset County Museum.	11.
3.	Ditto; small dorsal fin-spine —Ibid. B. M. no. 33476.	11.
4, 5.	*Hybodus strictus*, Agassiz; two dorsal fin-spines, the first exhibiting growth-lines.—Ibid. B. M. nos. 28447, P. 2835.	13.
6.	*Hybodus basanus*, Egerton (?); dorsal fin-spine.—Wadhurst Clay; Ecclesbourne, near Hastings. Rufford Collection (B. M. no. P. 8938).	10.
7.	Ditto; dorsal fin-spine.—Wealden; Hastings. Rufford Collection (B. M. no. P. 6936).	10.
8.	*Hybodus striatulus*, Agassiz; part of distal half of spine.—Tunbridge Wells Sands; Tilgate Forest. Mantell Collection (B. M. no. 2686).	13.

All the figures are of the natural size. The outlines 1 a–8 a represent transverse sections of the spines at the points marked by cross-lines.

PALÆONTOGRAPHICAL SOCIETY, 1915.

A.S.Woodward, Wealden & Purbeck Fishes. Plate III.

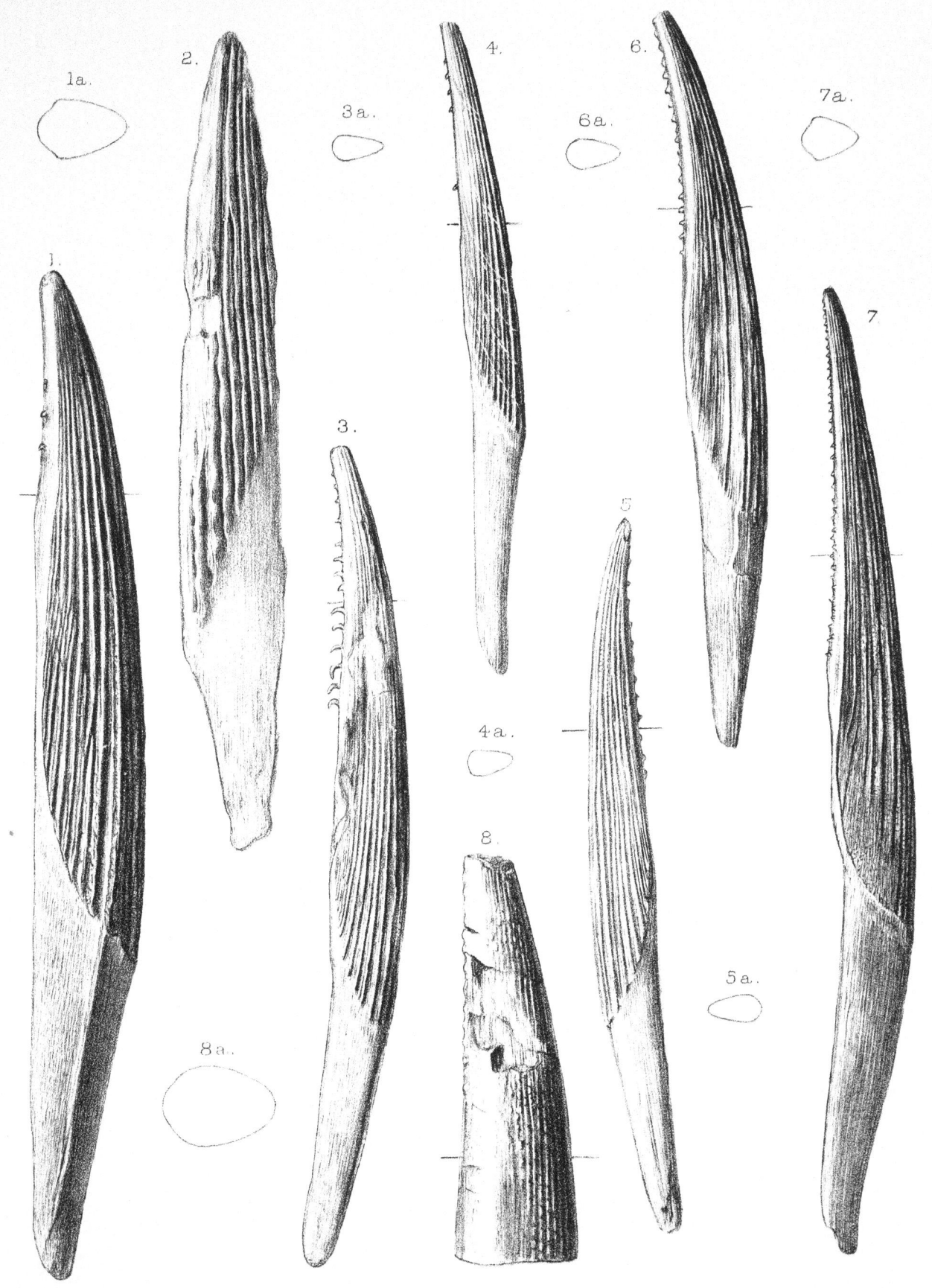

C.M.Woodward del. et lith. Huth imp.

Hybodus.

PLATE IV.

Fig.   Page.

1. *Undina purbeckensis*, sp. nov.; imperfect fish, one-half nat. size, with dorsal scales (1 *a*), ventral scales (1 *b*), and caudal flank scales (1 *c*) enlarged three times.—Middle Purbeck Beds; Swanage, Dorset. The type specimen. B. M. no. P. 11925. *a.*, fragment of anal fin; d^1, d^2, remains of two dorsal fins; *plv.*, base of pelvic fin.   22.

2. *Coccolepis andrewsi*, A. S. Woodward; fish wanting pectoral fins, three-halves nat. size.—Lower Purbeck Beds; Teffont, Wiltshire. The type specimen. Museum of Practical Geology, London, no. 419.   24.

3. Ditto; hinder half of fish, three-halves nat. size, with anal scale (3 *a*) enlarged ten times.—Ibid. B. M. no. P. 6302.   25.

4. *Coccolepis* sp.; imperfect right maxilla, outer view, twice nat. size. Wadhurst Clay; Buckshole Quarry, Silverhill, Hastings. Charles Dawson Collection (B. M. no P. 11924).   25.

PALÆONTOGRAPHICAL SOCIETY, 1915.

Plate IV.

A.S. Woodward, Wealden & Purbeck Fishes.

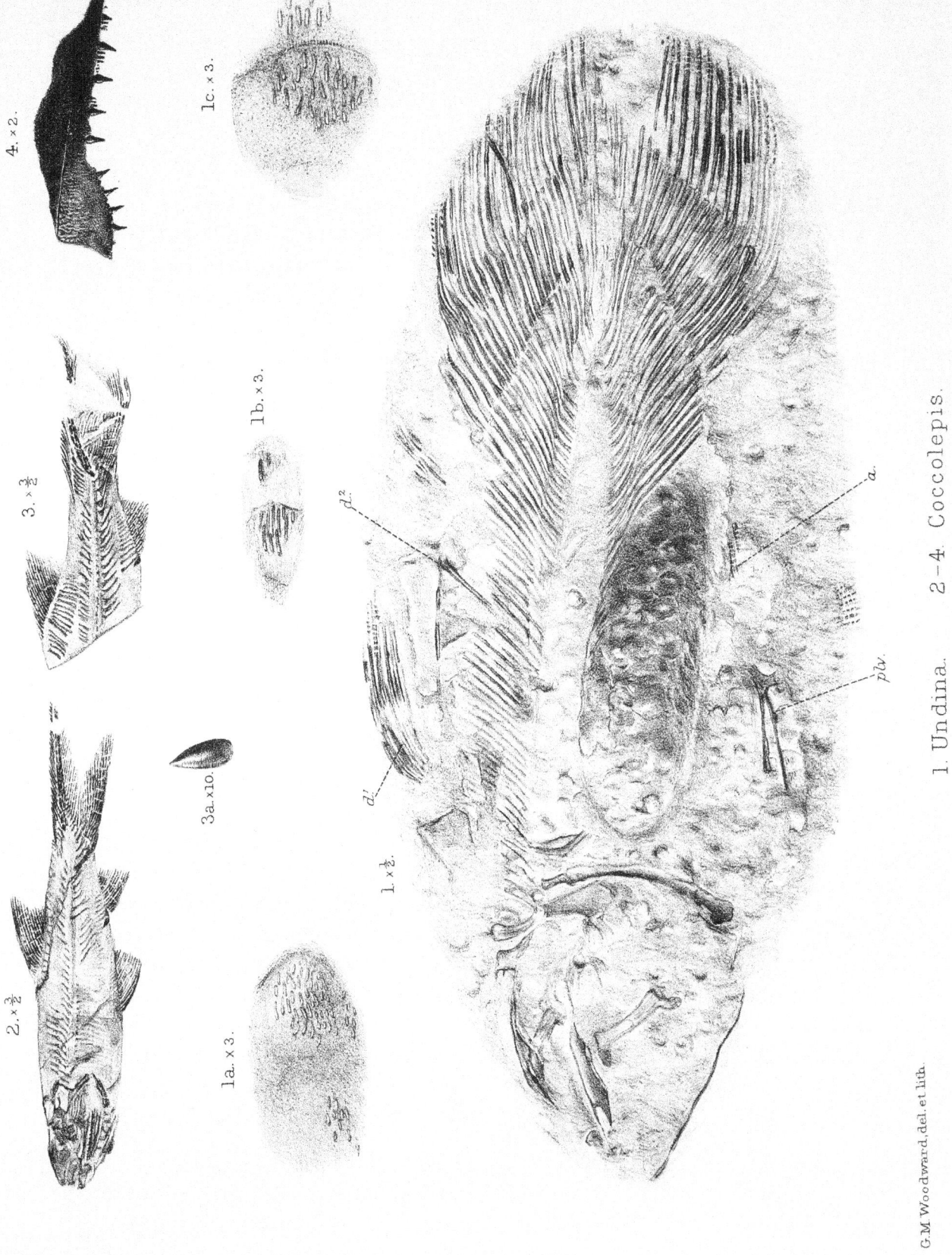

G.M Woodward, del. et lith.

Huth imp.

1. Undina. 2–4. Coccolepis.

PLATE V.

FIG.   PAGE.

1. *Hylæobatis problematica*, gen. et sp. nov.; crown of tooth from above and below (1 a), anterior view (1 b), hinder view (1 c), and two end views (1 d, 1 e), three times nat. size.—Wealden; Brook, Isle of Wight. The type specimen. York Museum.   19.

2. Ditto; half-worn crown of tooth, in upper and anterior view (2 a), three times nat. size.—Ibid. York Museum.   20.

3. Ditto; crown of large tooth, upper view and transverse section (3 a), three times nat. size.—Wealden; Sevenoaks, Kent. Sedgwick Museum, Cambridge.   20.

4, 5. Ditto; two worn teeth, upper view, three times nat. size.—Ibid. Sedgwick Museum, Cambridge.   21.

6. *Lepidotus minor*, Agassiz; inner view of vertically crushed head and anterior scales, nat. size.—Middle Purbeck Beds; Swanage, Dorset. Egerton Collection (B. M. no. P. 1118). *co.*, circumorbitals; *fr.*, frontal; *md.*, mandible showing splenial teeth; *op.*, operculum; *orb.*, orbit; *pa.*, parietal; *pmx*, premaxilla; *po.*, postorbital; *pop.*, pre-operculum; *ptt.*, post-temporal; *scl.*, supraclavicle; *sq.*, squamosal; *st.*, supratemporal.   29.

7. Ditto; head in upper and left side view, nat. size.—Ibid. Cunnington Collection (B. M. no. 36080). Lettering as in fig. 6.   29.

8. Ditto; imperfect head with base of pectoral fin and some anterior scales, left side view, nat. size.—Ibid. Museum of Practical Geology, London, no. 27974. *ag.*, angular; *br.*, branchiostegal rays; *ch.*, ceratohyal; *d.*, dentary; *iop.*, interoperculum; *pcl.*, postclavicular scales; *pro.*, preorbitals; *sop.*, suboperculum; other letters as in fig. 6.   29.

9. Ditto; parasphenoid, nat. size.—Ibid. Egerton Collection (B. M. no. P. 1121 a).   29.

10. Ditto; left hyomandibular, outer view, nat. size.—Ibid. B. M. no. 44848.   30.

11. Ditto; left premaxilla, outer view, nat. size.—Ibid. Beckles Collection (B. M. no. 48255).   30.

PALÆONTOGRAPHICAL SOCIETY, 1915.

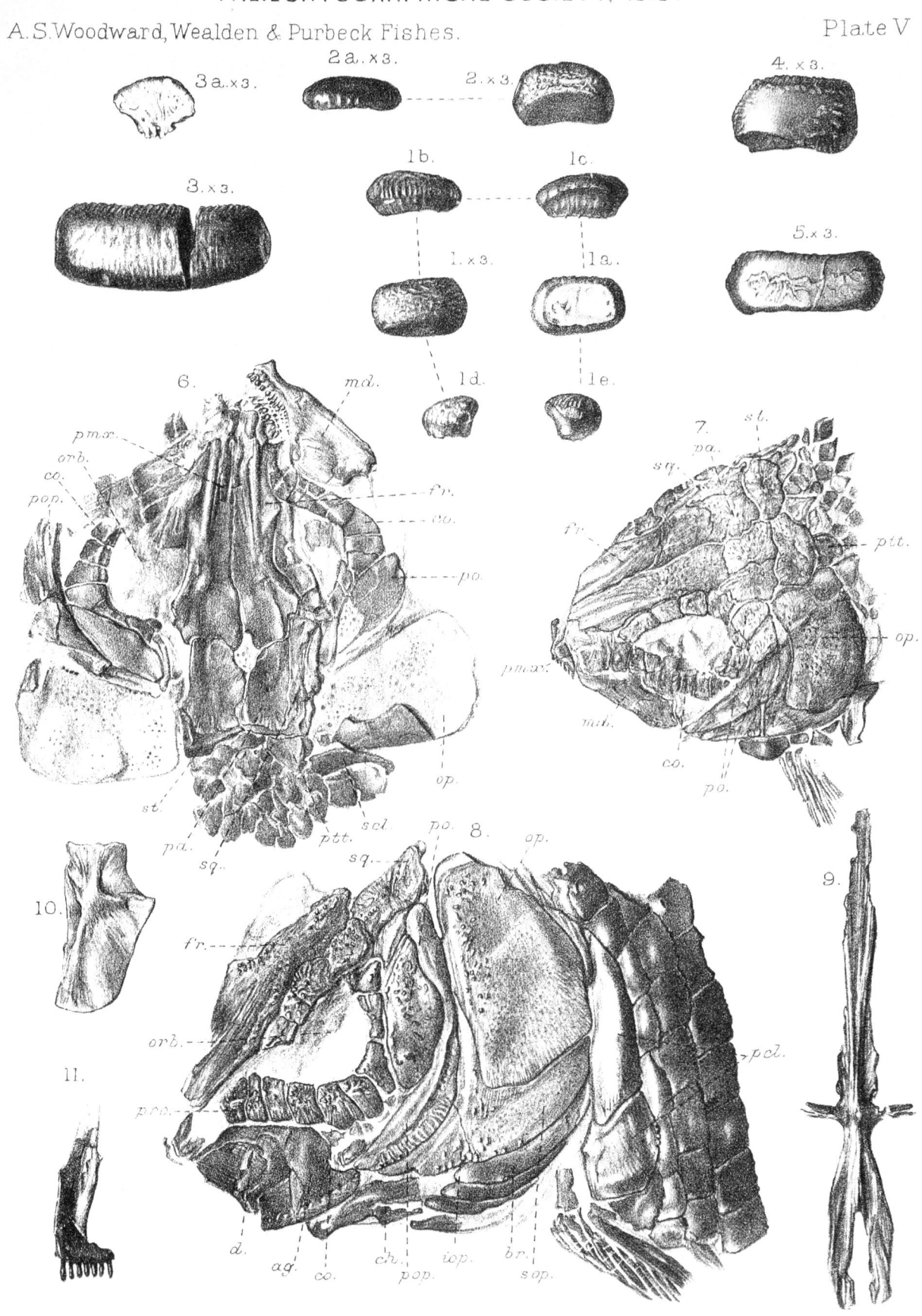

A. S. Woodward, Wealden & Purbeck Fishes.

Plate V

1–5. Hylæobatis.   6–11. Lepidotus.

PLATE VI.

Fig.
1. *Lepidotus minor*, Agassiz; nearly complete fish, left side view, nat. size, the left side of the head slightly displaced upwards, the ventral region of the trunk crushed downwards, and the dorsal ridge-scales lacking.—Middle Purbeck Beds; Swanage, Dorset. Museum of Practical Geology, London, no. 27975. *br.*, branchiostegal rays; *co.*, circumorbitals; *iop.*, interoperculum; *md.*, two rami of mandible, the left crushed upwards; *mx.*, maxilla, capped by supramaxilla; *orb.*, orbit; *pmx.*, premaxilla; *po.*, postorbital; *pop.*, preoperculum; *pro.*, preorbitals; *sop.*, suboperculum. 28.

PALÆONTOGRAPHICAL SOCIETY, 1915.　　　　　Plate VI.

A.S. Woodward, Wealden & Purbeck Fishes.

Lepidotus

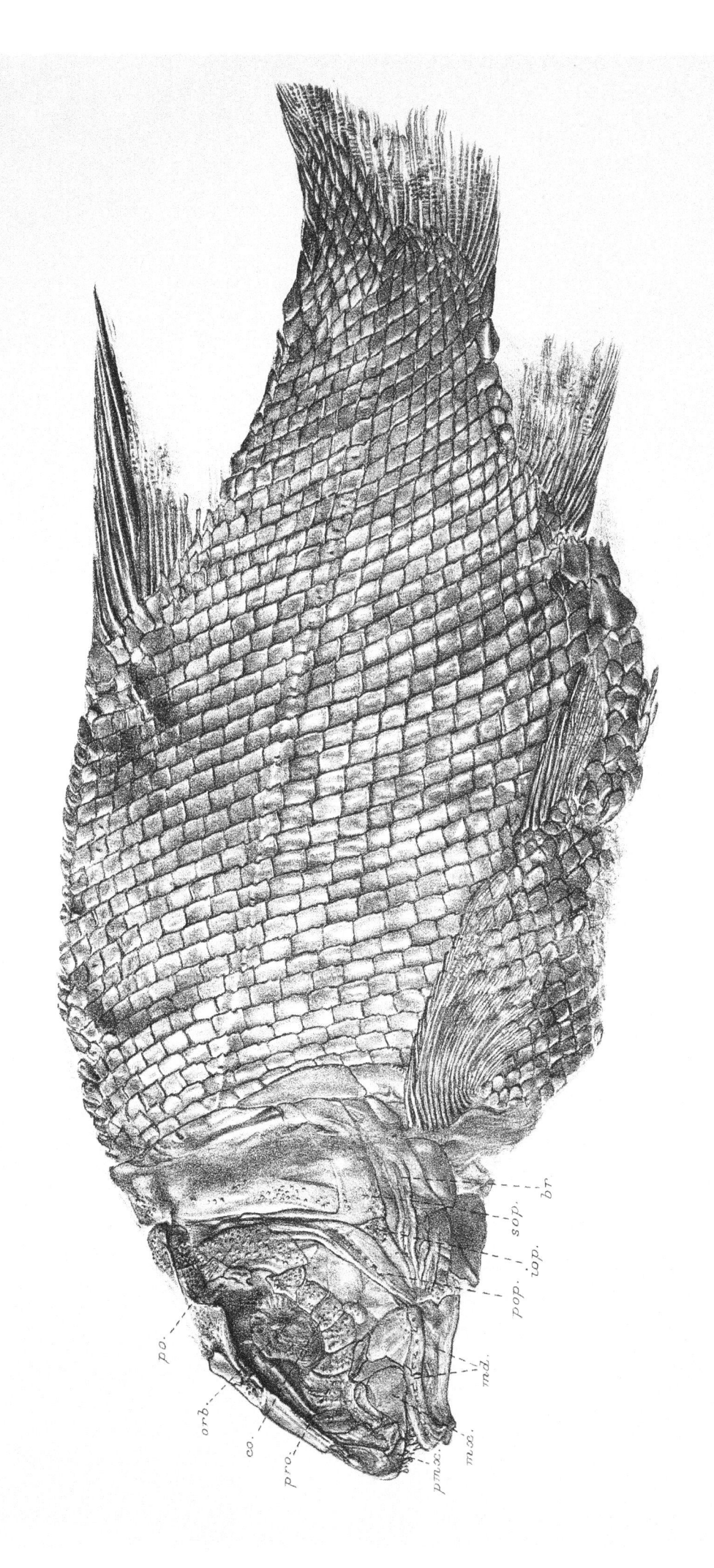

G.M. Woodward del et lith.　　　　　Huth imp.

PLATE VII.

Fig.
1. *Lepidotus minor*, Agassiz; left ectopterygoid, upper and outer (1 a) view, nat. size.—Middle Purbeck Beds; Swanage, Dorset. B. M. no. 21349.   30.
2. Ditto; right dentary, nat. size.—Ibid. B. M. no. 21974 a.   31.
3. Ditto; portion of trunk showing neural spines (*n.*) and ribs (*r*), nat. size.—Ibid. B. M. no. 45903.   32.
4. Ditto; dorsal ridge-scales, nat. size.—Ibid. Egerton Collection (B. M. no. P. 2006).   34.
5. Ditto; inner view of three flank-scales of left side, with inner view of abdominal ventral scales (5 *b*) and base of anal fin (5 *a*), nat. size.—Ibid. B. M. no. 41157. *f. s.*, endoskeletal support of fulcrated front border of anal fin.   33, 34.
6. *Lepidotus notopterus*, Agassiz; imperfect trunk, right side view, one-half nat. size.—Ibid. Enniskillen Collection (B. M. no. P. 4220).   35.
7. *Lepidotus mantelli*, Agassiz; imperfect fish, left side view, one-sixth nat. size.—Wadhurst Clay; Hastings. Charles Dawson Collection (B. M. no. P. 11832).   37.

PALÆONTOGRAPHICAL SOCIETY, 1915.

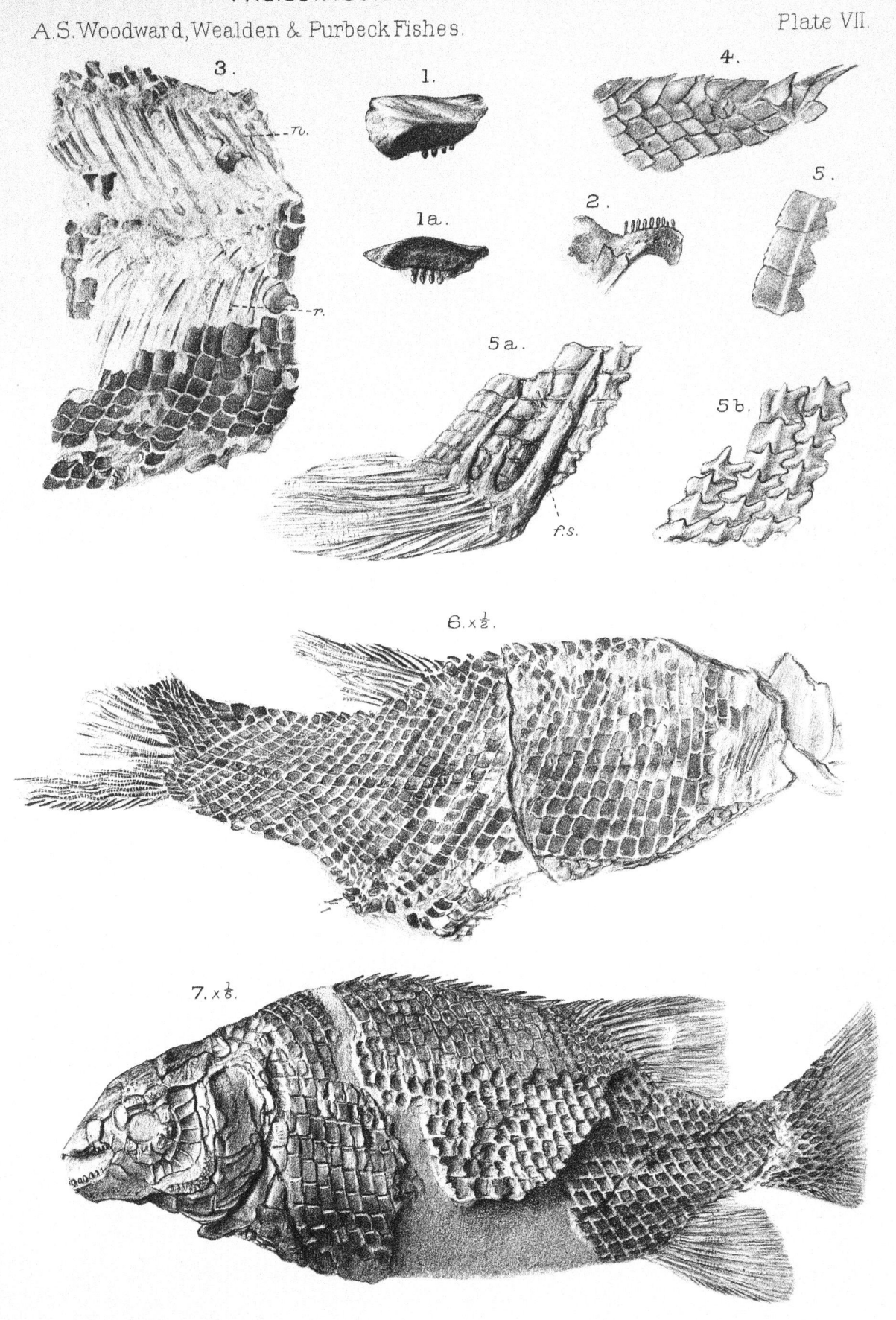

A.S.Woodward, Wealden & Purbeck Fishes.

Plate VII.

G.M.Woodward del.et lith.

Huth imp.

Lepidotus.

PLATE VIII.

FIG.   PAGE.

1. *Lepidotus mantelli*, Agassiz; hinder portion of head, with anterior scales and base of pectoral fin, one-half nat. size.—Wealden; Heathfield, Sussex. The type specimen. Mantell Collection (B. M. no. 2456). *cl.*, clavicle; *cor.*, coracoid; *hm.*, hyomandibular; *iop.*, interoperculum; *mpt.*, metapterygoid; *op.*, operculum; *pcl.*, postclavicular scale; *pct.*, pectoral fin; *pop.*, preoperculum; *scl.*, supraclavicle; *sop.*, suboperculum.   37.

2. Ditto; vertically crushed head in right side view and upper view (2 a), with a lower view of the dentary bones (2 b), one-half nat. size.—Wealden; Highfure, Billingshurst, Sussex. The type specimen of the so-called *Lepidotus fittoni*, Agassiz. B. M. no. 20673 a. *co.*, circumorbitals; *fr.*, frontal; *md.*, mandible; *pa.*, parietal; *po.*, postorbitals; *ptt.*, post-temporal; *sq.*, squamosal; *st.*, supratemporals; other letters as in fig. 1.   37.

3. Ditto; hinder margin of skull and anterior dorsal scales, upper view, one-half nat. size.—Wealden; Sussex. Mantell Collection (B. M. no. 2401). Lettering as in fig. 2.   44.

4. Ditto; flank scales, partially decayed, showing coarse oblique grooving, nat. size.—Wealden; Horsham, Sussex. B. M. no. P. 5129.   46.

PALÆONTOGRAPHICAL SOCIETY, 1915.

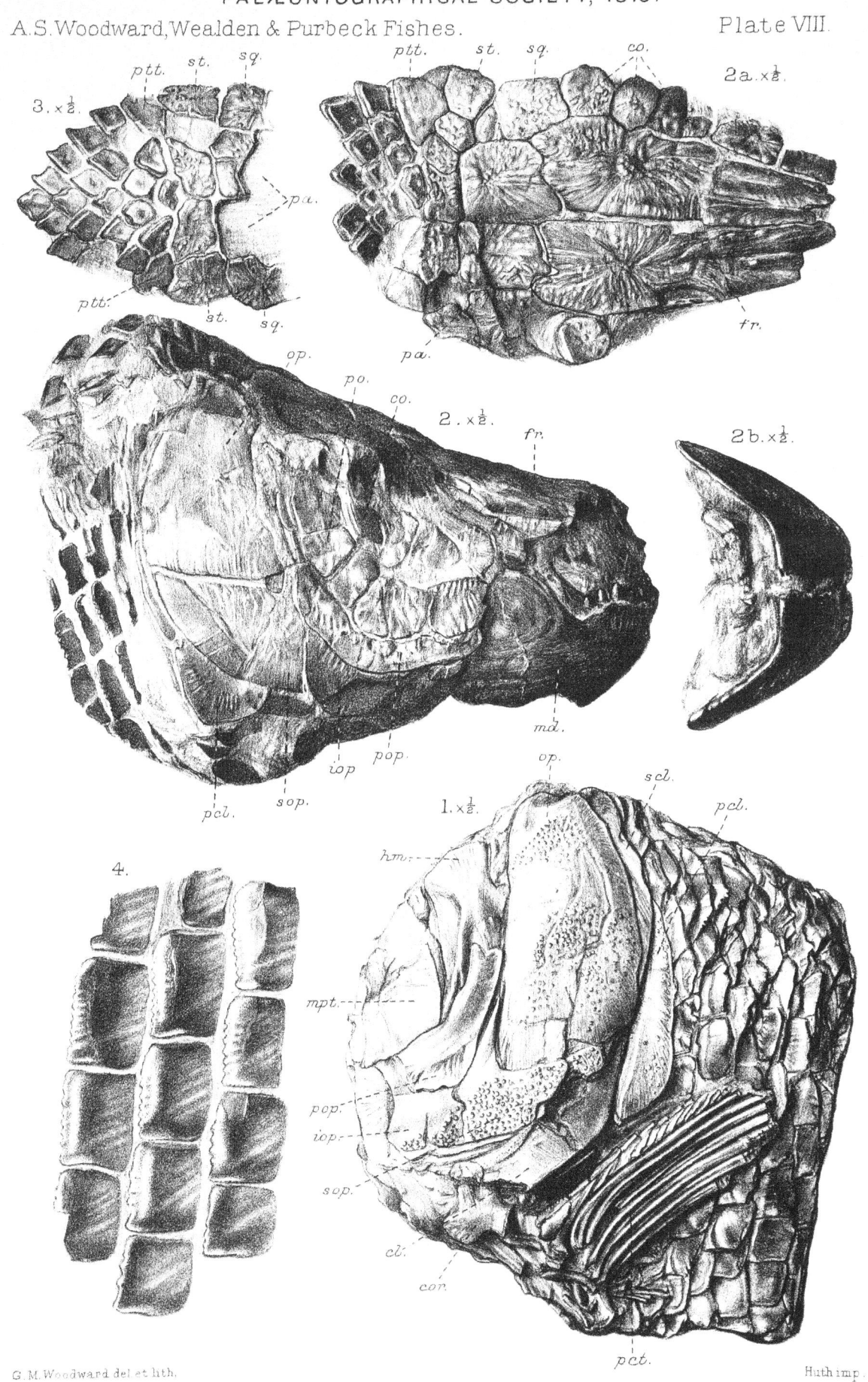

A. S. Woodward, Wealden & Purbeck Fishes.

Plate VIII.

Lepidotus.

G. M. Woodward del. et lith.

Huth imp.

PLATE IX.

FIG.
1. *Lepidotus mantelli*, Agassiz; small head and anterior scales, left side view, nat. size, with bones of cranial roof in outline (1 a), one-half nat. size.—Wealden; Hastings. Rufford Collection (B. M. no. P. 6933). *ag.*, angular; *cl.*, clavicle; *co.*, circumorbitals; *d.*, dentary; *fr.*, frontal; *iop.*, interoperculum; *l.*, scales of lateral line; *mx.*, maxilla; *op.*, operculum; *pa.*, parietal; *pcl.*, postclavicular scale; *po.*, postorbitals (the large foremost plate of the series lacking); *pop.*, preoperculum; *pro.*, preorbitals; *ptt.*, post-temporal; *scl.*, supraclavicle; *smx.*, supramaxilla; *sop.*, suboperculum; *sq.*, squamosal; *st.*, supratemporal.   39.

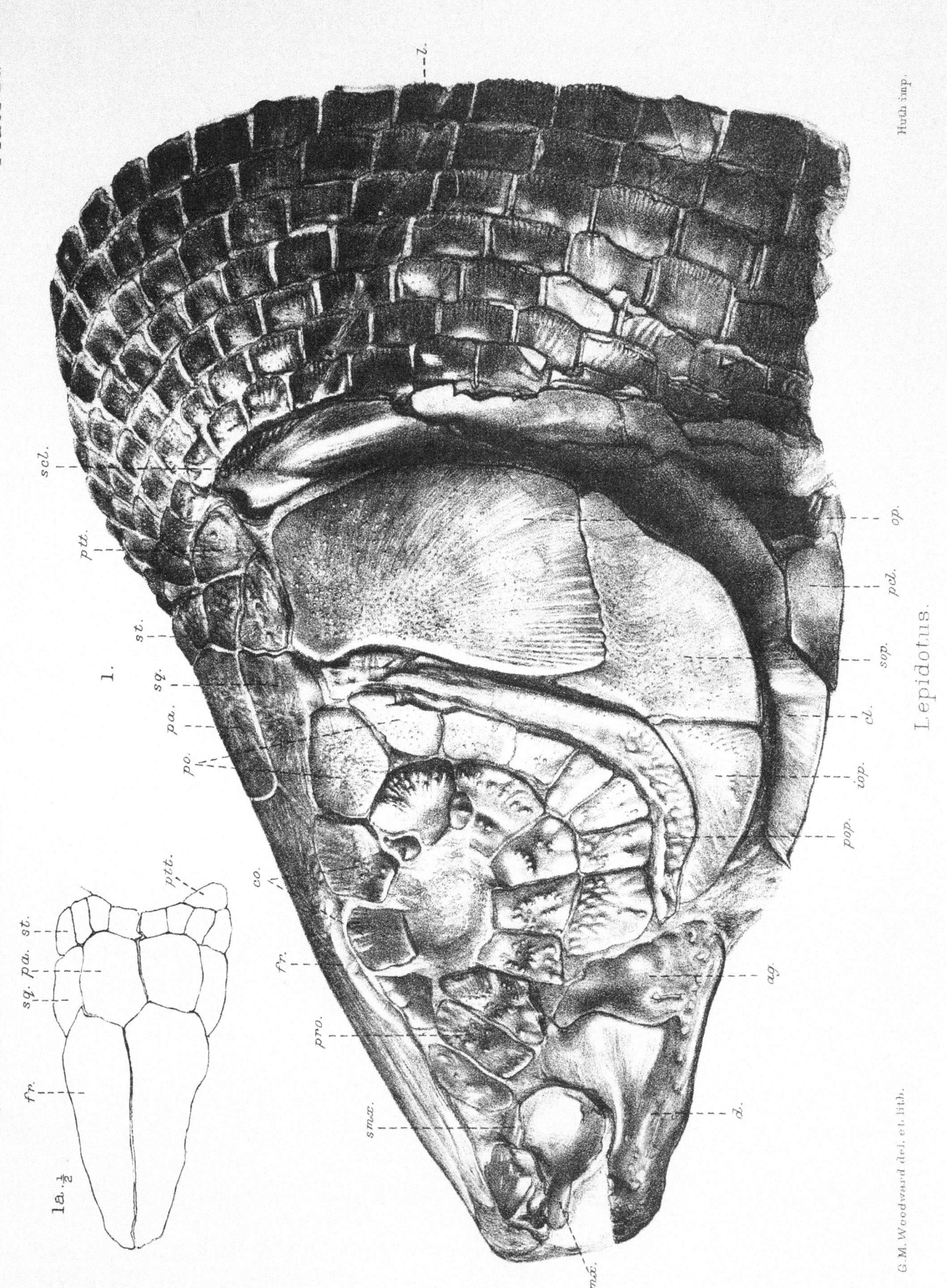

Plate IX.

A.S. Woodward, Wealden & Purbeck Fishes

Lepidotus.

PLATE X.

Fig.   Page.

1. *Lepidotus mantelli*, Agassiz; occipital portion of skull, abraded below, hinder view and left side view (1 a), nat. size.—Wealden; Hastings. Egerton Collection (B. M. no. P. 1124). *bo.*, fragment of basi-occipital; *epo.*, epiotic; *exo.*, exoccipital; *f. m.*, foramen magnum; *opo.*, opisthotic; *pro.*, pro-otic; *st.*, supratemporals; x., foramen for exit of vagus nerve.   37.

2. Ditto; hyoid arch, two-thirds nat. size.—Upper Purbeck Beds; Perch Hill, Brightling, Sussex. B. M. no. 23624. *br.*, upper branchiostegal ray; *ch.*, ceratohyal; *eph.*, epihyal; *hyh.*, hypohyal.   43.

3. Ditto; head in upper and right side view, and remains of jaws in front view (3 a), nat. size —Upper Purbeck Beds; Netherfield, Battle, Sussex. E. J. Baily Collection (Hastings Museum). *ag.*, angular; *co.*, circumorbitals; *d.*, dentary; *fr.*, frontal; *iop.*, interoperculum; *mx.*, maxilla; *op.*, operculum; *pa.*, parietal; *pmx.*, premaxilla; *po.*, postorbitals; *pop.*, preoperculum; *pro.*, preorbitals; *ptt.*, post-temporal; *sop.*, suboperculum; *sq.*, squamosal; *st.*, supratemporals.   37.

PALÆONTOGRAPHICAL SOCIETY, 1915.

A. S. Woodward, Wealden & Purbeck Fishes.

Plate X.

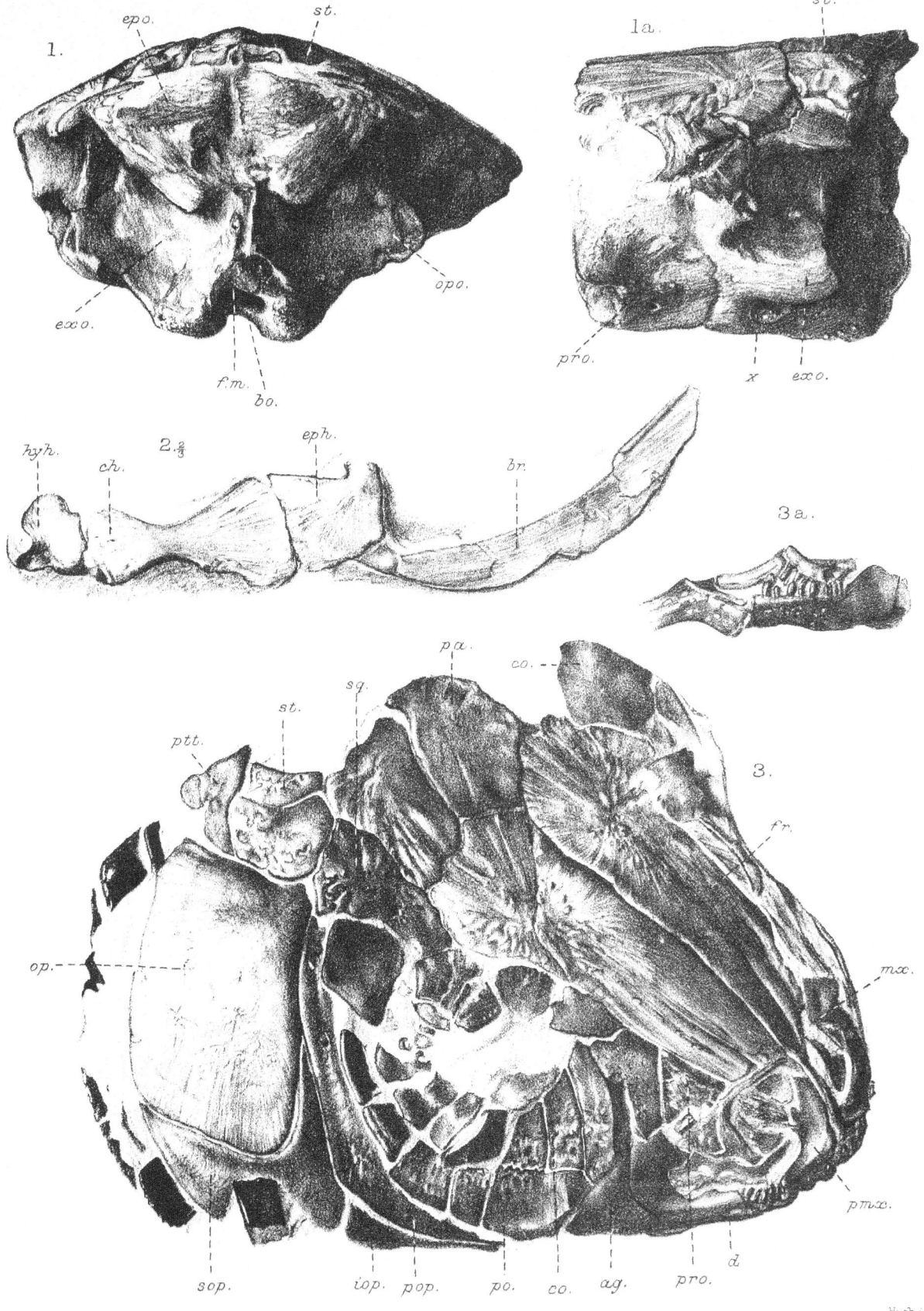

Lepidotus.

PLATE XI.

FIG.		PAGE.
1.	*Lepidotus mantelli*, Agassiz; anterior end of vomer (*v.*) and left pterygo-palatine (*pt.*) in cross-section, anterior view, to show successional teeth in sockets, nat. size.—Wealden; Hastings. Beckles Collection (B. M. no. P. 6342).	42.
2.	Ditto; vomer, oral, anterior (2 *a*), and right lateral (2 *b*) views, nat. size, with three anterior teeth (2 *c*), enlarged twice to show mammilliform apex.—Ibid. Beckles Collection (B. M. no. P. 6363).	42.
3.	Ditto; right pterygo-palatine, oral, outer (3 *a*), and inner (3 *b*) views, nat. size; *s.*, successional teeth in sockets.—Ibid. Dawson Collection (B. M. no. P. 4917).	41.
4.	Ditto; vomerine (*v.*) and right pterygo-palatine (*pt.*) dentition, oral view, nat. size, with two teeth in side view (4 *a*, 4 *b*), enlarged twice to show wrinkled gano-dentine.—Wealden; Isle of Wight. British Museum, no. 47504.	43.
5.	Ditto; imperfect right splenial (*spl.*) and dentary (*d.*), oral, outer (5 *a*), and inner (5 *b*) views, showing successional teeth (*s.*) in broken section of bone, nat. size.—Wealden; Tilgate Forest. Mantell Collection, no. 2326.	42.
6.	Ditto; three teeth of right dentary bone, outer view, three times nat. size.—Wealden; Hastings. J. E. Lee Collection (B. M. no. P. 4995).	42.
7.	Ditto; middle flank-scale, outer view, nat. size.—Ibid. Dawson Collection (B. M. no. P. 4916).	47.
8.	Ditto; middle flank-scale, inner view, nat. size.—Wealden; Tilgate Forest. Mantell Collection, no. 2397.	47.
9, 10.	Ditto; upper flank-scales of abdominal region, outer view, and second in inner view (10 *a*), nat. size.—Ibid. Mantell Collection, nos. 3036, 3092.	47.
11, 12.	Ditto; lower flank-scales of abdominal region, outer view and inner view (11 *a*, 12 *a*), nat. size.—Ibid. Mantell Collection, nos. 3089, 3045.	47.
13.	Ditto; caudal scale, outer and inner view (13 *a*), nat. size.—Ibid. Mantell Collection (B. M. no. 26009).	47.
14.	Ditto; caudal scale of lateral line, outer and inner view (14 *a*), nat. size.—Ibid. Mantell Collection.	47.

PALÆONTOGRAPHICAL SOCIETY, 1916.

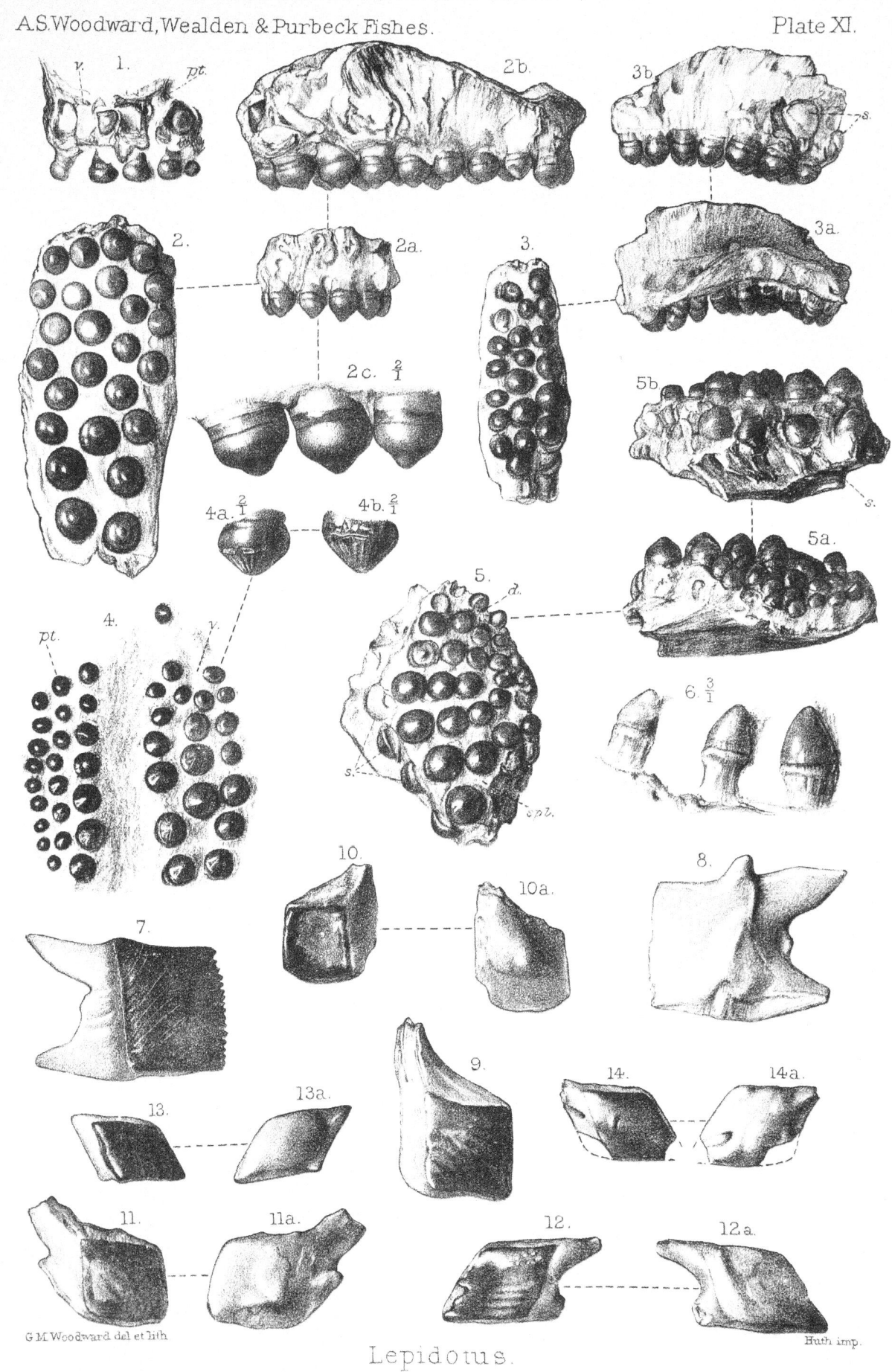

A.S.Woodward, Wealden & Purbeck Fishes.

Plate XI.

G.M.Woodward del et lith.

Huth imp.

Lepidotus.

PLATE XII.

FIG.   PAGE.

1. *Mesodon daviesi*, A. S. Woodward; nearly complete fish, two-thirds nat. size, with some splenial teeth (1 *a*), two vomerine teeth (1 *b*), other teeth from the counterpart (1 *c*), and the caudal scale (1 *d*), enlarged three times.—Middle Purbeck Beds; Swanage, Dorset. The type specimen. British Museum, nos. 41387, P. 6381. *op.*, operculum; *pop.*, preoperculum.   50.

2. Ditto; crushed head and anterior abdominal region, six-fifths nat. size.—Ibid. Enniskillen Collection (B. M. no. P. 4381). *cl.*, clavicle; *d.*, dentary; *op.*, operculum; *pop.*, preoperculum of both sides, the right from within, the left imperfect.   51.

3. *Mesodon parvus*, A. S. Woodward; fish, lacking head, three-halves nat. size.—Middle Purbeck Beds; Teffont, Wiltshire. The type specimen. Rev. W. R. Andrews Collection (B. M. no. P. 9845). *br.*, branchiostegal rays; *l.l.*, calcifications along lateral line; *op.*, operculum; *pop.*, preoperculum.   52.

4. Ditto; fish, lacking tail, three-halves nat. size, with part of mandible (4 *a*) showing dentary (*d.*) and basal rim of splenial teeth, enlarged six times.—Ibid. T. T. Gething Collection (B. M. no. P. 10954).   53.

PALÆONTOGRAPHICAL SOCIETY, 1916.

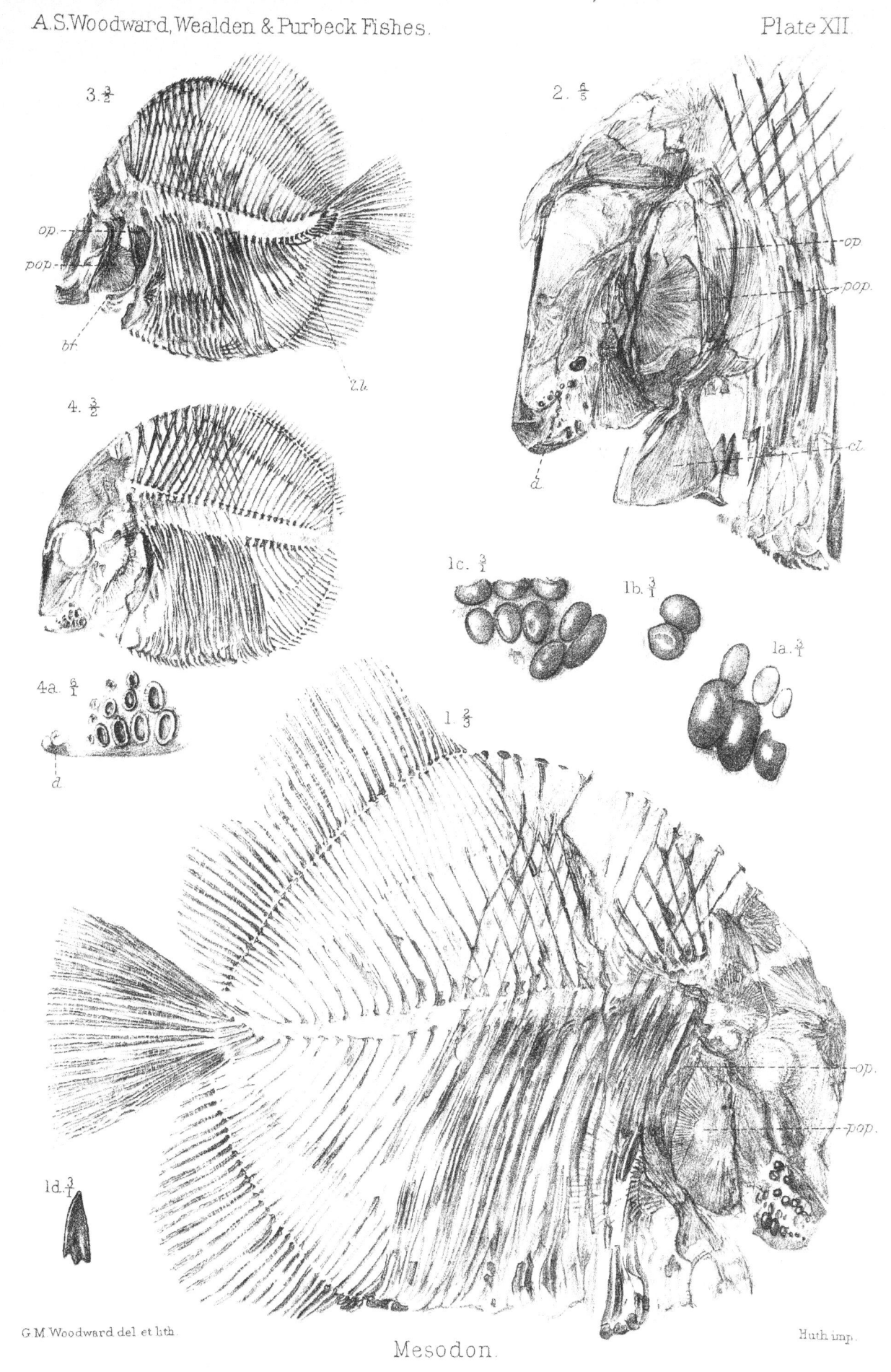

A.S. Woodward, Wealden & Purbeck Fishes.

Plate XII.

G.M. Woodward del et lith.

Huth imp.

Mesodon.

PLATE XIII.

FIG.   PAGE.

1. *Eomesodon barnesi*, A. S. Woodward; imperfect fish, nat. size, with denticles of dorsal ridge-scales (1 *a*, 1 *b*) enlarged three times, and ventral ridge-scales (1 *c*) with anterior anal fin-rays (*a*.) enlarged twice.—Middle Purbeck Beds; Swanage, Dorset. Beckles Collection (B. M. no. P. 6382).   57.

2. *Eomesodon depressus*, sp. nov.; imperfect fish, two-thirds nat. size, with a dorsal ridge-scale (2 *a*) enlarged twice, two flank-scales, outer view (2 *b*) and inner view (2 *c*), three-halves nat. size.—Ibid. The type specimen. British Museum, no. P. 10583. *d.*, supports of dorsal fin; *orb.*, orbit.   57.

3. *Mesodon* sp.; left splenial dentition, oral view, twice nat. size.—Ibid. Enniskillen Collection (B. M. no. P. 3757).   58.

4. *Mesodon* sp.; left splenial dentition, oral view, twice nat. size.—Ibid. Museum of Practical Geology, London, no. 28355.   58.

5. *Cœlodus arcuatus*, sp. nov.; vomerine dentition, oral view, twice nat. size.—Ibid. Museum of Practical Geology, London, no. 28353.   70.

PALÆONTOGRAPHICAL SOCIETY, 1916.

A.S.Woodward, Wealden & Purbeck Fishes.

Plate XIII.

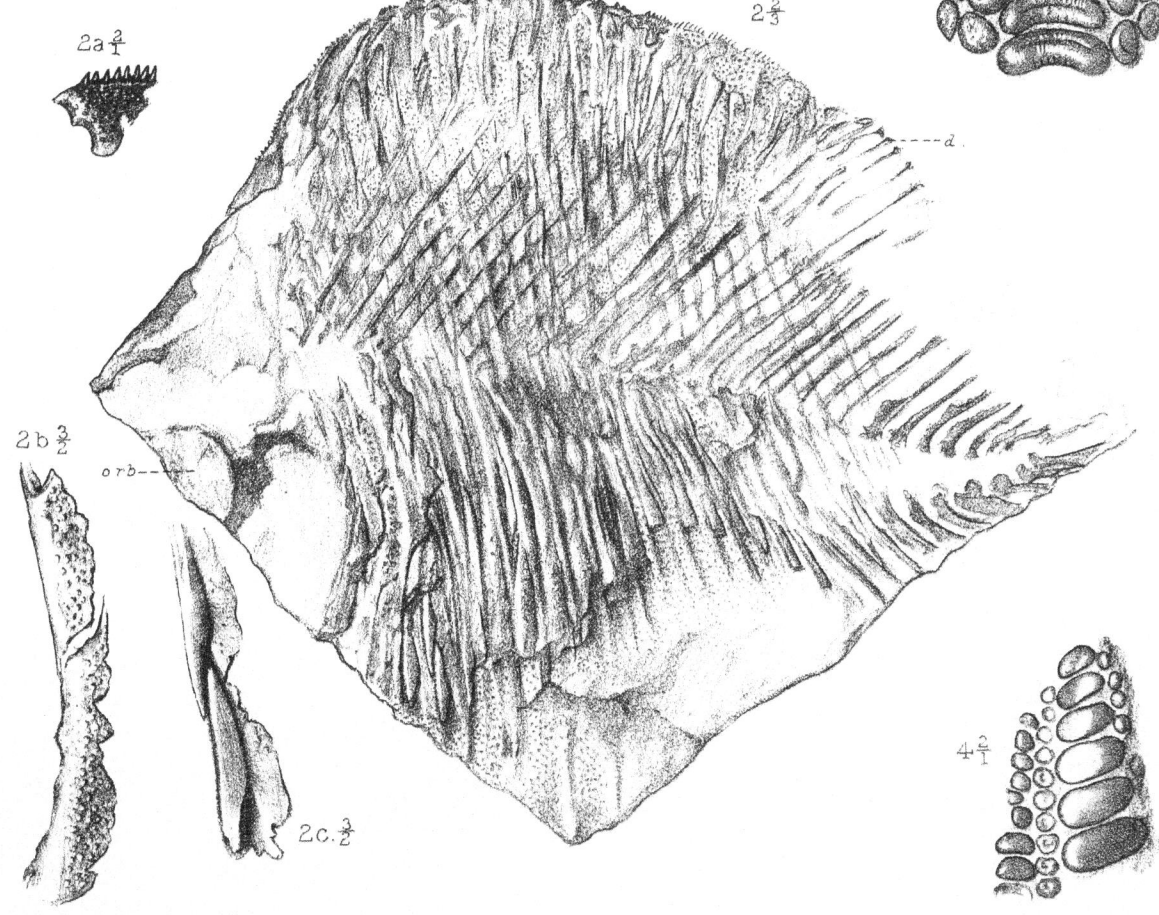

G.M.Woodward del et lith.

Huth.imp.

1, 2. Eomesodon.   3, 4. Mesodon.   5 Cœlodus.

PLATE XIV.

Fig.		Page.
1.	*Microdon radiatus*, Agassiz; nearly complete fish, nat. size, with dorsal (1 a) and ventral (1 b) ridge-scale enlarged three times.—Middle Purbeck Beds; Swanage, Dorset. Dorset County Museum, Dorchester. *x.*, problematical bone at hinder end of abdominal region.	59.
2.	Ditto; imperfect fish with distorted tail, nat. size.—Ibid. Cunnington Collection (B. M. no. 46333). *c.*, displaced caudal fin-rays; *cl.*, clavicle; *dr.*, ventral ridge-scale with double articulation; *fr.*, frontal bone; *me.*, mesethmoid; *plv.*, pelvic fin; *vo.*, vomer.	59.
3.	Ditto; head and anterior abdominal region, nat. size.—Ibid. Dorset County Museum, Dorchester. *ag.*, angular; *d.*, dentary; *hm.*, displaced hyomandibular; *r.*, two expanded anterior ribs; *spl.*, splenial.	61.
4.	Ditto; left parietal bone (drawn upside down), showing hinder prominence (*x.*), twice nat. size.—Ibid. Egerton Collection (B. M. no. P. 1627 a).	60.
5.	Ditto; vomerine dentition, three times nat. size.—Ibid. Museum of Practical Geology, London, no. 28359.	62.
6.	Ditto; right dentary bone, five times nat. size.—Ibid. Museum of Practical Geology, London, no. 28352.	61.
7.	Ditto; splenials with dentition, oral view, with some lateral vomerine teeth to the left, three times nat. size.—Ibid. British Museum, no. P. 5592.	62.
8.	Ditto; right splenial dentition, oral view, three times nat. size.—Ibid. Egerton Collection (B. M. no. P. 1627 a).	62.
9.	Ditto; epihyal (*eph.*) and ceratohyal (*ch.*) bones, five times nat. size.—Ibid. Cunnington Collection (B. M. no. 28443).	62.
10.	Ditto; left hyomandibular (*hm.*), outer view, with remains of operculum (*op.*) partially overlapping the supraclavicle (*scl.*), twice nat. size.—Ibid. Egerton Collection (B. M. no P. 1627 b).	61.

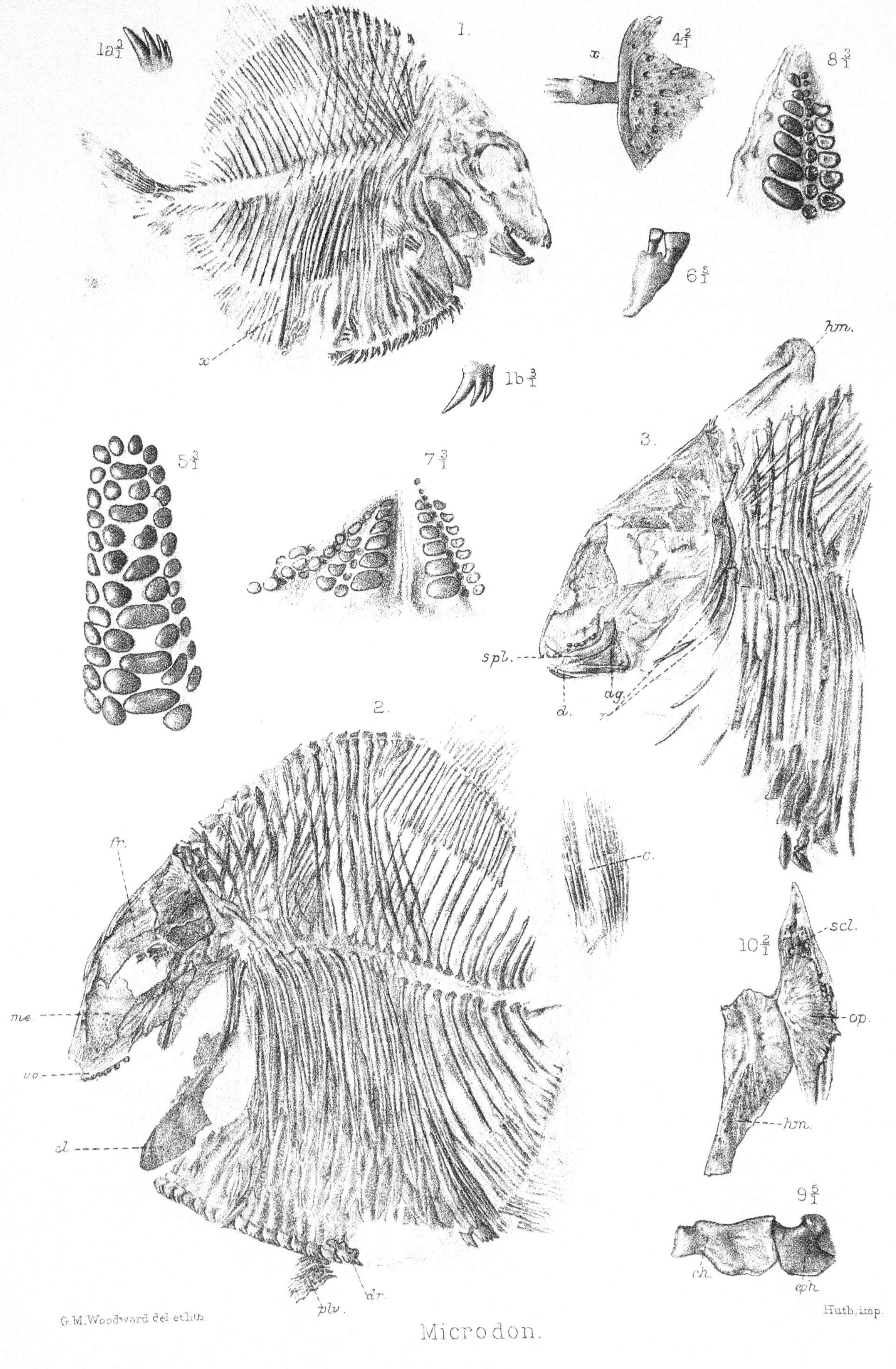

PALÆONTOGRAPHICAL SOCIETY, 1916.

A. S. Woodward. Wealden & Purbeck Fishes.

Plate XIV

G. M. Woodward del. et lith.

Huth, imp.

Microdon.

PLATE XV.

Fig.		Page
1.	*Microdon radiatus*, Agassiz; skull, right side view, twice nat. size.—Middle Purbeck Beds; Swanage, Dorset. British Museum, no. P. 7455. *fr.*, frontal; *k.*, lower median keel on parasphenoid; *me.*, mesethmoid; *ors.*, orbitosphenoid (?); *pa.*, parietal; *pas.*, parasphenoid; *r.s.*, ridge-scale fused with occiput; *socc.*, supraoccipital; *sq.*, squamosal; *vo.*, vomer.	59.
2.	Ditto; caudal fin, twice nat. size.—Ibid. British Museum, no. 19013.	63.
3.	Ditto; some abdominal flank-scales, left side, outer view, twice nat. size.—Ibid. Cunnington Collection (B. M. no. 46333).	64.
4.	Ditto; some abdominal flank-scales, right side, inner view showing riblets, twice nat. size.—Ibid. British Museum, no. 44844.	64.
5, 5 a.	Ditto; two abdominal flank-scales, right side, inner view, showing peg-and-socket articulation, twice nat. size.—Ibid. Dorset County Museum, Dorchester.	64.
6.	*Cœlodus mantelli*, Agassiz; portion of vomerine dentition, oral view and right side view (6 a), three-halves nat. size, with a median tooth enlarged three times (6 b), and a transverse section of the dentition (6 c), three-halves nat. size.—Wealden; Tilgate Forest, Sussex. Mantell Collection (B. M. no. 28417).	66.
7.	Ditto; right splenial, oral view, three-halves nat. size.—Wealden; Hastings, Sussex. Enniskillen Collection (B. M. no. P. 3763).	67.
8.	Ditto; portion of right splenial dentition, oral view, twice nat. size.—Wealden; Tilgate Forest. Mantell Collection, no. 2709.	67.
9.	Ditto; portion of left splenial dentition, oral view, three times nat. size.—Ibid. Type specimen of *Gyrodus mantelli*, Agassiz. Mantell Collection (B. M. no. 28415 a).	67.
10.	Ditto; left splenial, oral view, twice nat. size.—Ibid. Mantell Collection (B. M. no. 28415 b).	67.
11.	Ditto; left splenial, oral view, and hinder view (11 a), twice nat. size.—Lower Wealden or Upper Purbeck Beds; Netherfield, Sussex. Teilhard & Pelletier Collection (B. M. no. P. 11903).	67.
12.	*Cœlodus multidens*, sp. nov.; right splenial, oral view, twice nat. size.—Wealden; Sevenoaks, Kent. The type specimen. Museum of Practical Geology, London, no. 7492.	68.
13.	Ditto; portion of right splenial dentition, oral view, twice nat. size.—Wealden; Battle, Sussex. Bowerbank Collection (B. M. no. 39215).	68.
14.	*Cœlodus hirudo* (Agassiz); a principal tooth of the left splenial dentition, oral view, front view (14 a), and end view (14 b), three-halves nat. size.—Wealden; Tilgate Forest, Sussex. The type specimen. Mantell Collection, no. 2706.	68.
15.	Ditto; a principal tooth probably of the vomerine dentition, oral view, hinder view (15 a), and lower view (15 b), three-halves nat. size.—Wealden; Hastings, Sussex. Rufford Collection (B. M. no. P. 9838).	68.
16.	Ditto; a principal tooth, oral view, twice nat. size.—Wealden (Wadhurst Clay); Hastings. Teilhard & Pelletier Collection (B. M. no. P. 11900).	69.
17.	Ditto; an imperfect lateral tooth, three times nat. size.—Ibid. Teilhard & Pelletier Collection (B. M. no. P. 11901).	69.
18.	Ditto; lateral tooth, five times nat. size.—Ibid. Teilhard & Pelletier Collection (B. M. no. P. 11902).	69.
19.	*Cœlodus lævidens*, sp. nov.; right splenial, oral view, three-halves nat. size.—Middle Purbeck Beds; Swanage, Dorset. The type specimen. British Museum, no. P. 10679.	69.
20.	Ditto; left splenial, oral view, twice nat. size.—Ibid. British Museum, no. 33480.	69.

PALÆONTOGRAPHICAL SOCIETY, 1916.

A. S. Woodward, Wealden & Purbeck Fishes.

Plate XV

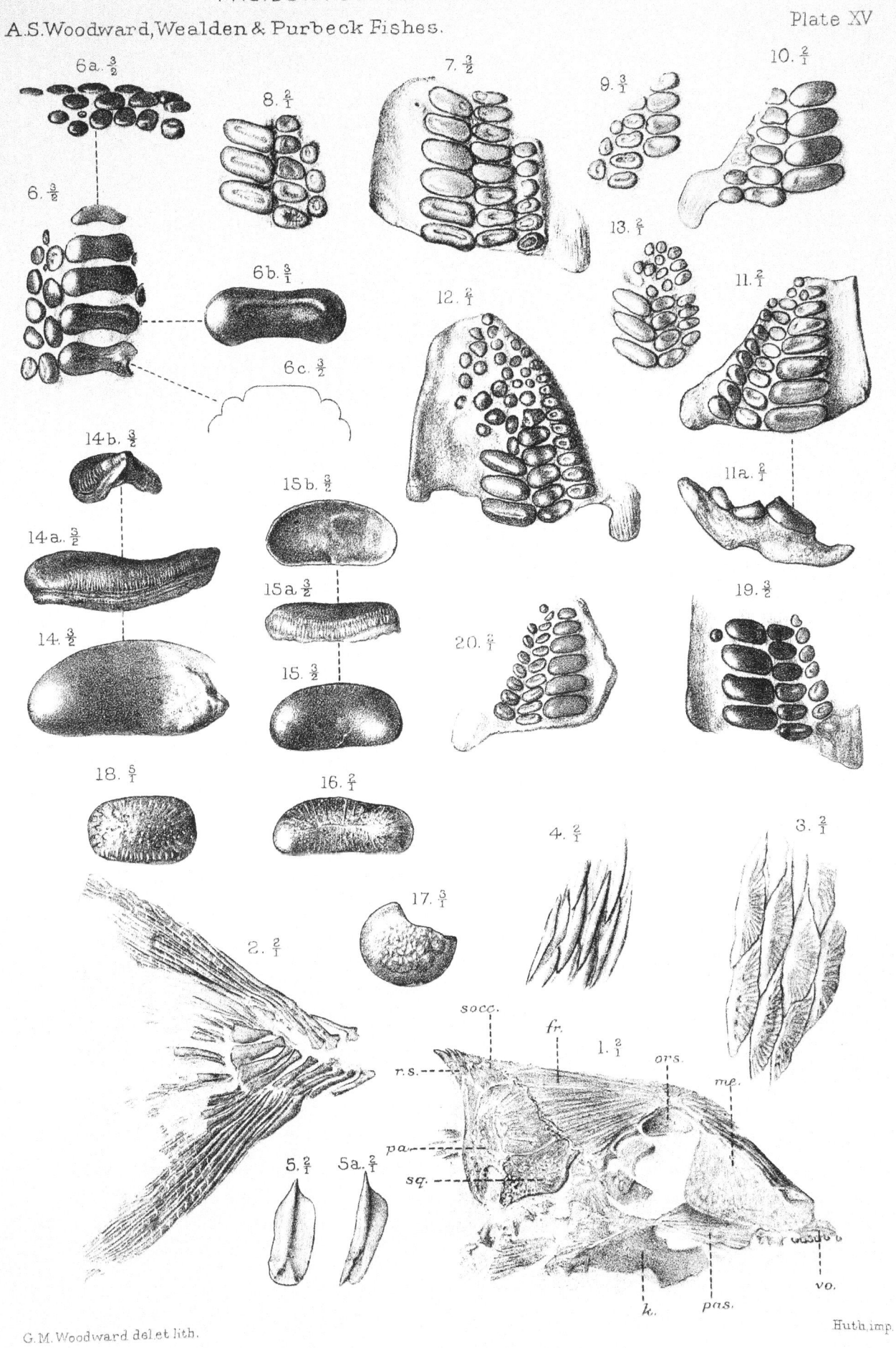

G. M. Woodward del. et lith.   Huth, imp.

1–5. Microdon.   6–20. Cœlodus.

PLATE XVI.

FIG.   PAGE.

1. *Ophiopsis penicillata*, Agassiz; nearly complete fish, showing most of squamation from inner face, nat. size, with a group of teeth (1 *a*), some abdominal flank-scales, inner view (1 *b*), and some caudal scales, outer view (1 *c*), enlarged five times.—Probably from Lower Purbeck Beds near Weymouth, Dorset. The type specimen. British Museum, no. P. 7433.   72.

2. Ditto; bases of some dorsal fin-rays, five times nat. size.—Lower Purbeck Beds; Isle of Portland. British Museum, no. P. 8375.   73.

3. *Ophiopsis breviceps*, Egerton; nearly complete fish, nat. size.—Lower Purbeck Beds; Wockley, near Tisbury, Wiltshire. The type specimen. Museum of Practical Geology, London, no. 28442.   74.

4. Ditto; roof of skull and some adjacent bones, inner view, five-halves nat. size.—Ibid. British Museum, no. P. 9107 *a*. *co.*, upper circumorbitals; *fr.*, frontal; *pa.*, parietal; *ptt.*, post-temporal; *sq.*, squamosal; *st.*, supratemporal.   74.

5. Ditto; right premaxilla, outer view, four times nat. size.—Ibid. British Museum, no. P. 9436.   74.

6. Ditto; right mandibular ramus, outer view, four times nat. size.—Ibid. British Museum, no. P. 9436.   74.

7. Ditto; portion of operculum to show ornament, four times nat. size.—Ibid. Enniskillen Collection (B. M. no. P. 3608).   74.

8. Ditto; left supraclavicle, outer view, four times nat. size.—Ibid. British Museum, no. P. 9436.   75.

9, 10. Ditto; flank-scales, outer and inner face, seven and eight times nat. size.—Ibid. British Museum, nos. P. 9107 *b, c*.   75.

11. Ditto; ventral scale, outer face, six times nat. size.—Ibid. British Museum, no. P. 9436.   75.

12. Ditto; dorsal scale, inner face, eight times nat. size.—Ibid. British Museum, no. P. 9107 *b*.   75.

13. *Ophiopsis dorsalis*, Agassiz; nearly complete fish, nat. size.—Probably from Middle Purbeck Beds, Swanage, Dorset. The type specimen. Egerton Collection (B. M. no. P. 466).   76.

PALÆONTOGRAPHICAL SOCIETY, 1916.

A.S.Woodward, Wealden & Purbeck Fishes.

Plate XVI.

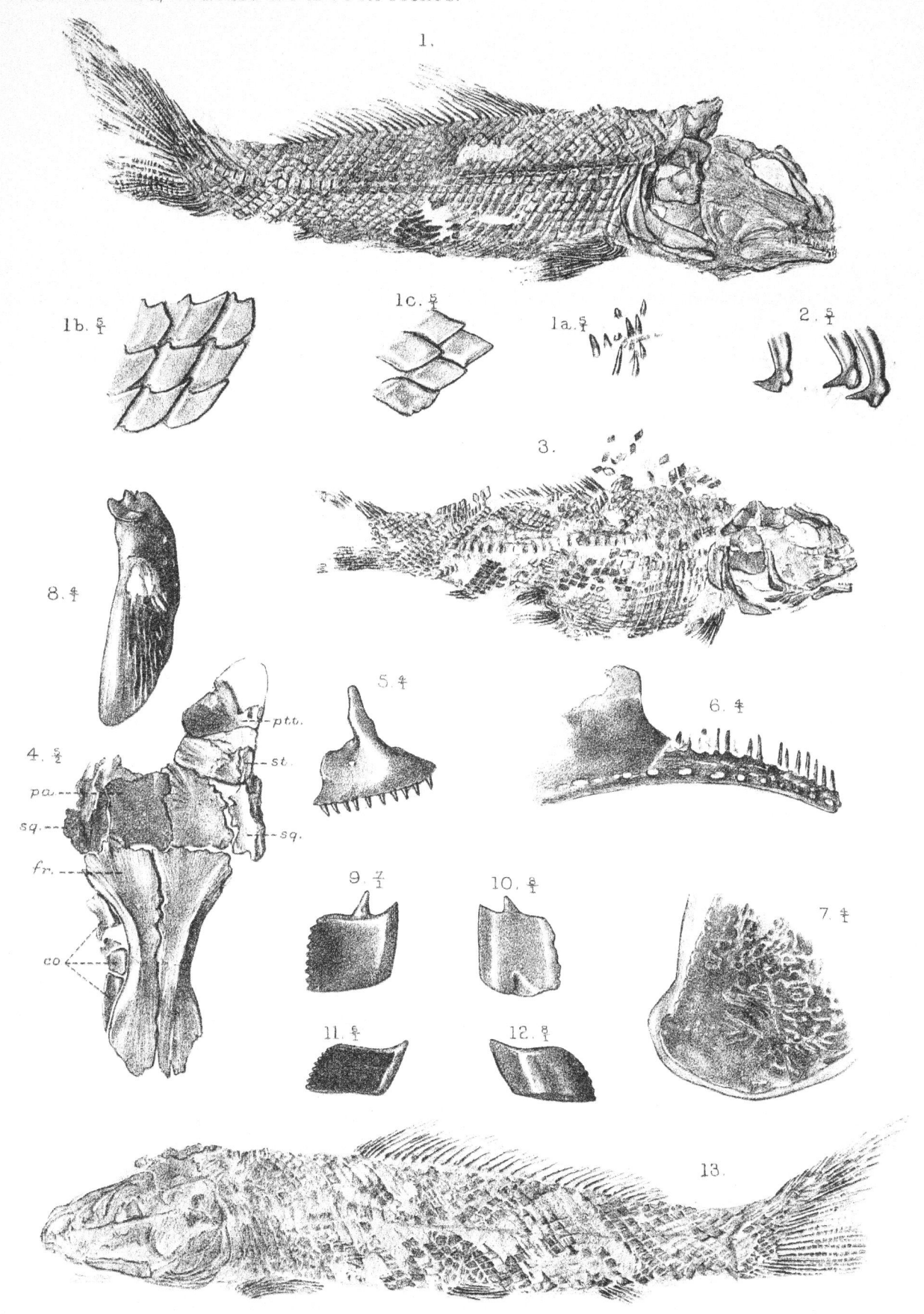

G.M.Woodward del. et lith.

Huth imp

Ophiopsis.

PLATE XVII.

FIG.		PAGE.
1.	*Histionotus angularis*, Egerton; fish with imperfect dorsal fin, lacking greater part of caudal region, nat. size.—Middle Purbeck Beds; Swanage. The type specimen. Egerton Collection (B. M. no. P. 577). *cl.*, clavicle; *d.*, dentary; *mx.*, maxilla; *pa.*, parietal; *pop.*, preoperculum; *ptt.*, post-temporal; *scl.*, supraclavicle; *st.*, supratemporal.	77.
2.	Ditto; nearly complete fish, with crushed cranium and distorted tail, nat. size.—Ibid. Dorset County Museum, Dorchester. *br.*, branchiostegal rays; *d.*, dentary; *iop.*, interoperculum; *mx.*, maxilla; *op.*, operculum; *pa.*, parietal; *pcl.*, post-clavicular scales; *pop.*, preoperculum; *sop.*, suboperculum.	77.
3.	Ditto; roof of skull, two-and-a-half times nat. size.—Ibid. British Museum, no. P. 5935. *co.*, upper circumorbitals; *fr.*, frontal; *pa.*, parietal.	77.
4.	Ditto. Supposed palatine or ectopterygoid, with comparatively stout teeth, three times nat. size.—Middle Purbeck Beds; Tisbury, Wiltshire. Cunnington Collection (B. M. no. 46421).	78.
5.	Ditto; inner view of three flank-scales, one traversed by the slime-canal of the lateral line (*l.*), two-and-a-half times nat. size, and (5 *a*) a flank-scale, outer face, three times nat. size.—Middle Purbeck Beds; Swanage. Enniskillen Collection (B. M. no. P. 3614).	79.
6.	*Enchelyolepis andrewsi*, A. S. Woodward; nearly complete fish, two-and-a-half times nat. size, with a dorsal fin-support (6 *a*) enlarged twelve times, and some scales (6 *b*) enlarged fifteen times.—Middle Purbeck Beds; Teffont, Wiltshire. The type specimen. Rev. W. R. Andrews Collection (B. M. no. P. 6303).	81.
7.	*Enchelyolepis pectoralis* (Sauvage); nearly complete fish, twice nat. size, with a dorsal fin-support (7 *a*) and a scale (7 *b*) enlarged ten times.—Upper Portlandian; Savonnières-en-Perthois, Meuse, France. The type specimen. British Museum, no. P. 7359.	81
8.	*Œonoscopus* sp.; right maxilla, outer view, nat. size.—Middle Purbeck Beds; Swanage. British Museum, no. 33477.	

PALÆONTOGRAPHICAL SOCIETY, 1916.

A. S. Woodward, Wealden & Purbeck Fishes.

Plate XVII.

G. M. Woodward del. et lith.

Huth imp.

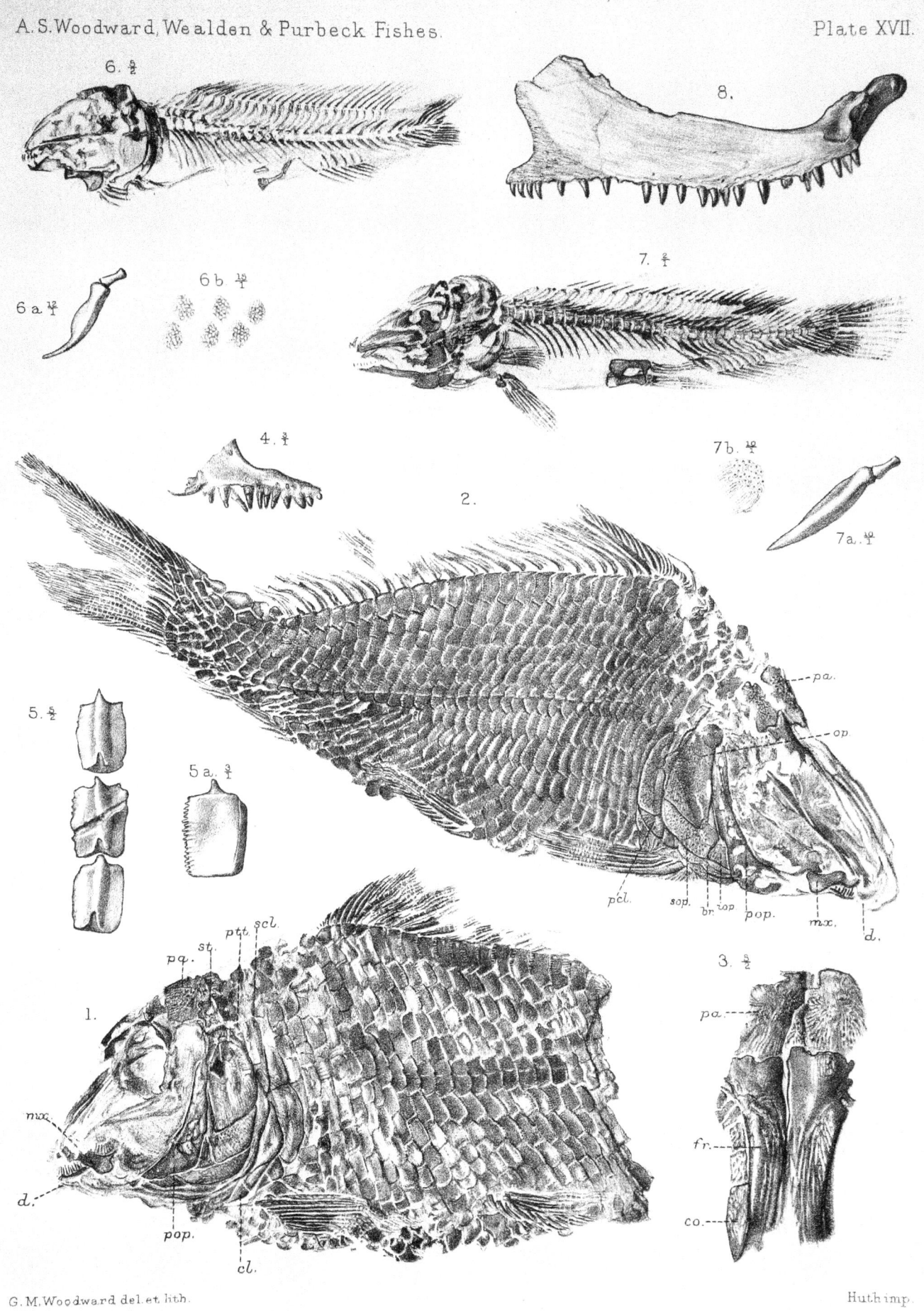

1–5. Histionotus. 6, 7. Enchelyolepis. 8. Oeonoscopus.

PLATE XVIII.

FIG.		PAGE
1.	*Neorhombolepis valdensis*, A. S. Woodward; imperfect fish coiled up in ironstone, nat. size.—Wealden; Hastings. The type specimen. Beckles Collection (B. M. no. P. 6364).	88.
2, 3.	Ditto; portions of head, pectoral arch, pectoral fin, and anterior scales of opposite side of same specimen, nat. size.	88.
4.	Ditto; remains of dentary bone of same specimen, nat. size.	88.

br., branchiostegal rays; *c.*, remains of caudal fin; *cl.*, clavicle; *co.*, circumorbital; *d.*, base of dorsal fin; *hm*, hyomandibular; *md.*, hinder end of mandible; *mx.*, portion of maxilla; *n.a.*, neural arches of vertebræ; *pa.*, parietal; *pas.*, parasphenoid; *pcl.*, postclavicular plates; *pct.*, pectoral fin; *po.*, portion of postorbital; *ptf.*, postfrontal (sphenotic); *qu.*, quadrate; *scl.*, supraclavicle; *sq.*, squamosal; *sy.*, symplectic; *v.*, vertebral centrum; *x.*, interoperculum (?)

PALÆONTOGRAPHICAL SOCIETY, 1916.

A. S. Woodward, Wealden & Purbeck Fishes.

Plate XVIII.

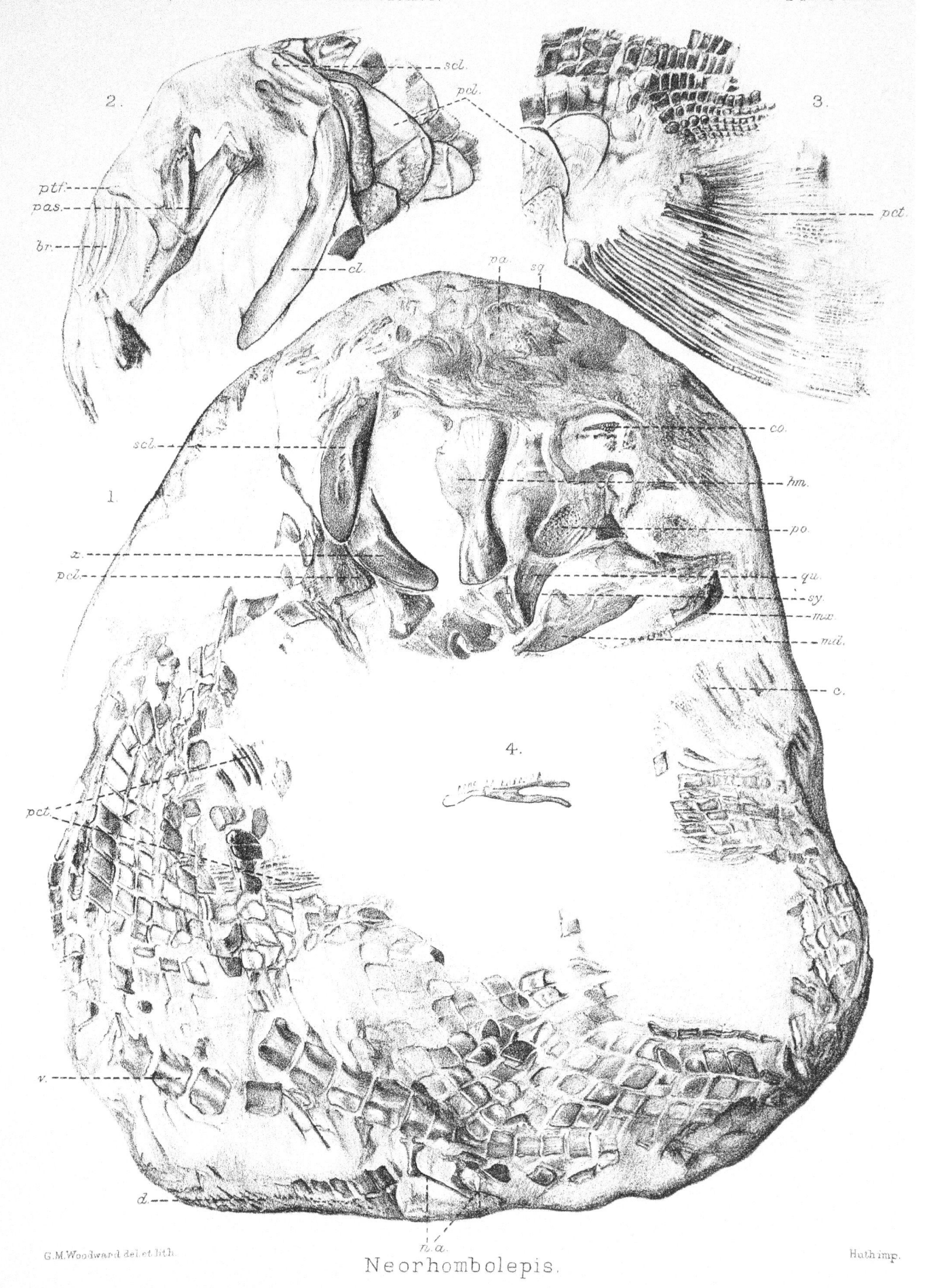

G. M. Woodward del. et lith.

Neorhombolepis.

Huth imp.

PLATE XIX.

Fig.		Page.
1.	*Caturus purbeckensis*, A. S. Woodward; head, etc., crushed in right side view, nat. size.—Middle Purbeck Beds; Swanage. The type specimen. British Museum, no. 46911. *cl.*, clavicle; *d.*, dentary; *hy.*, hypocentrum; *mx.*, maxilla; *pct.*, base of pectoral fin; *pmx.*, premaxilla; *smx.*, supramaxilla.	85.
2.	Ditto; left mandibular ramus, outer view, nat. size.—Ibid. Sedgwick Museum, Cambridge. *ag.*, angular; *co.*, coronoid; *d.*, dentary.	86.
3.	*Caturus tenuidens*, A. S. Woodward; left maxilla, inner view, nat. size, with two teeth enlarged five times (3 *a*).—Ibid. . Egerton Collection (B. M. no. P. 969).	87.
4.	Ditto; right dentary, outer view, nat. size, with two teeth enlarged four times (4 *a*).—Ibid. British Museum, no. 40656.	87.
5.	*Amiopsis damoni* (Egerton); nearly complete fish, right side view, nat. size —Purbeck Beds; Portland. British Museum, no. 41156.	91.
6.	Ditto; some abdominal flank-scales of type specimen, four times nat. size.—Purbeck Beds; Bincombe, near Weymouth. Egerton Collection (B. M. no. P. 563.)	92.
7.	*Amiopsis* sp.; right maxilla, without teeth, outer view, nat. size.—Lower Purbeck Beds; Portland. British Museum, no. P. 8376.	94.
8.	*Amiopsis* sp.; left mandibular ramus, inner view, nat. size, with points of teeth broken away.—Ibid. British Museum, no. P. 8377.	94.

PALÆONTOGRAPHICAL SOCIETY, 1916.

A. S. Woodward. Wealden & Purbeck Fishes.

Plate XIX.

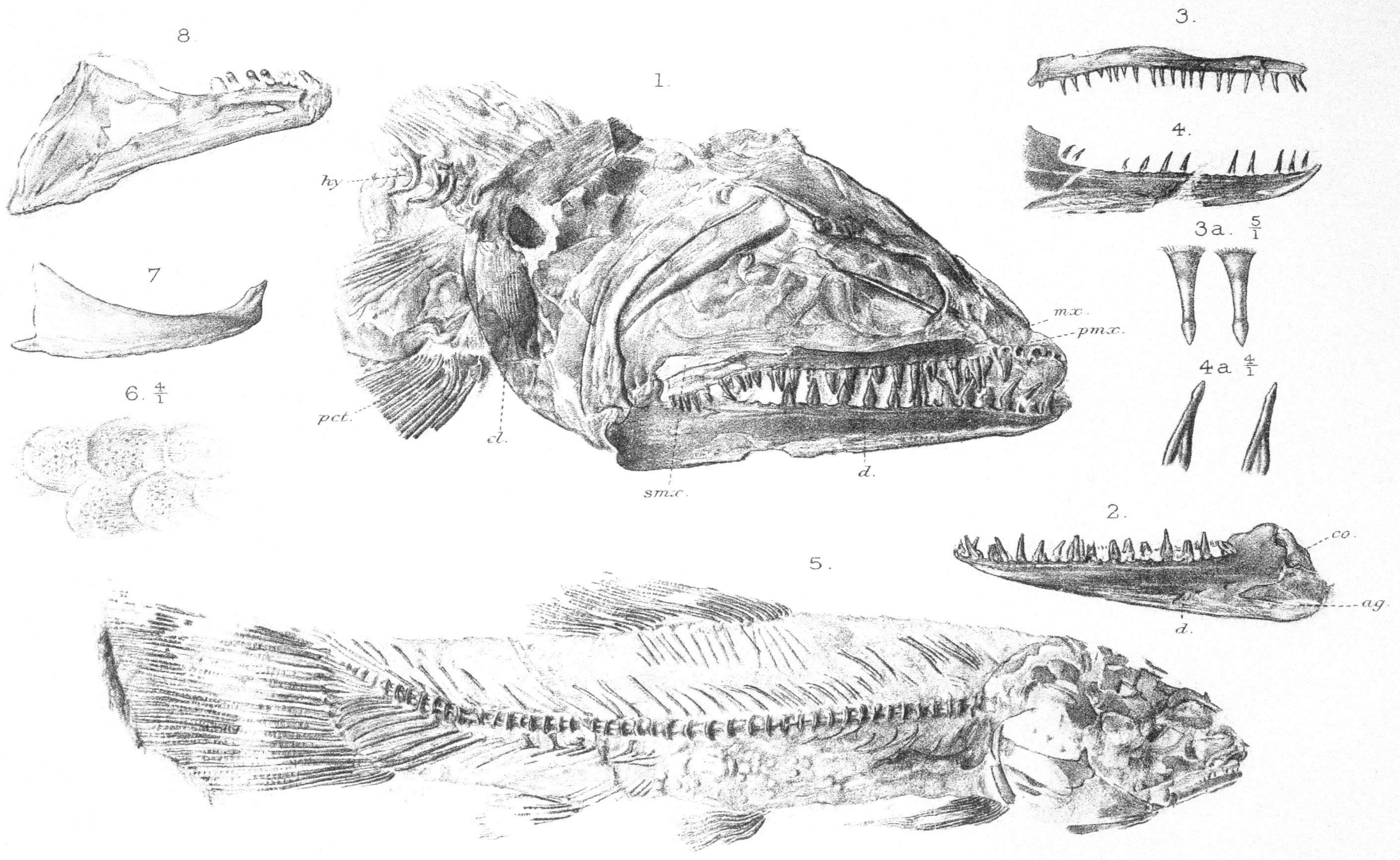

G. M. Woodward del. et lith.

Huth, imp.

1–4. Caturus.   5–8. Amiopsis.

PLATE XX.

FIG.   PAGE.

1. *Aspidorhynchus fisheri*, Egerton; almost complete fish, right side view, nearly two-thirds nat. size, with remains of branchial arches (1 a) and some dorsal caudal scales (1 b) enlarged three times.—Middle Purbeck Beds; Swanage. The type specimen. Dorset County Museum, Dorchester. *a.*, anal fin; *d.*, dorsal fin; *plv.*, pelvic fins.   97.

2. Ditto; head, etc., left side view and partly upper view, three-halves nat. size.—Ibid. British Museum, no. 28621. *ag.*, angular; *cl.*, clavicle; *d.*, dentary; *fr.*, frontal; *hm.*, hyomandibular; *mx.*, maxilla; *op.*, operculum; *pa.*, parietal; *pas.*, parasphenoid; *pct.*, pectoral fin; *pmx.*, premaxilla; *ps.*, presymphysial bone; *qu.*, quadrate; *smx.*, supramaxilla; *sy.*, symplectic; *x.*, probably epiotics.   97.

3. Ditto; portion of abdominal squamation, nat. size, with (3 a) three ventral scales five-halves nat. size.—Ibid. Museum of Practical Geology, London, no. 28440.   99.

4. Ditto; dorsal caudal scales, three times nat. size.—Ibid. Beckles Collection (B. M. no. P. 6380).   99.

5. *Pholidophorus ornatus*, Agassiz; imperfect tail, nat. size, with scale (5 a) enlarged twice.—Ibid. The type specimen. Mantell Collection (B. M. no. P. 7583). *a.*, anal fin; *plv.*, pelvic fin; *v.*, vertebral centrum.   102.

6. Ditto; small fish, left side view, nat. size, with (6 a) jaws enlarged twice, and (6 b) abdominal flank-scales enlarged three times.—Ibid. Alexander J. Hogg Collection (B. M. no. P. 10011). *ag.*, angular; *d.*, dentary; *mx.*, maxilla; *pmx.*, premaxilla; *smx.*, supramaxilla.   102.

7. Ditto; caudal scales of lateral line, outer view, five-halves nat. size, and abdominal ventral scales (7 a) of same specimen, inner view, three times nat. size.—Ibid. British Museum, no. 43038.   105.

8. Ditto; abdominal ventral scales, outer view, three times nat. size.—Ibid. Museum of Practical Geology, London, no. 28439.   105.

PALÆONTOGRAPHICAL SOCIETY, 1916.

Plate XX.

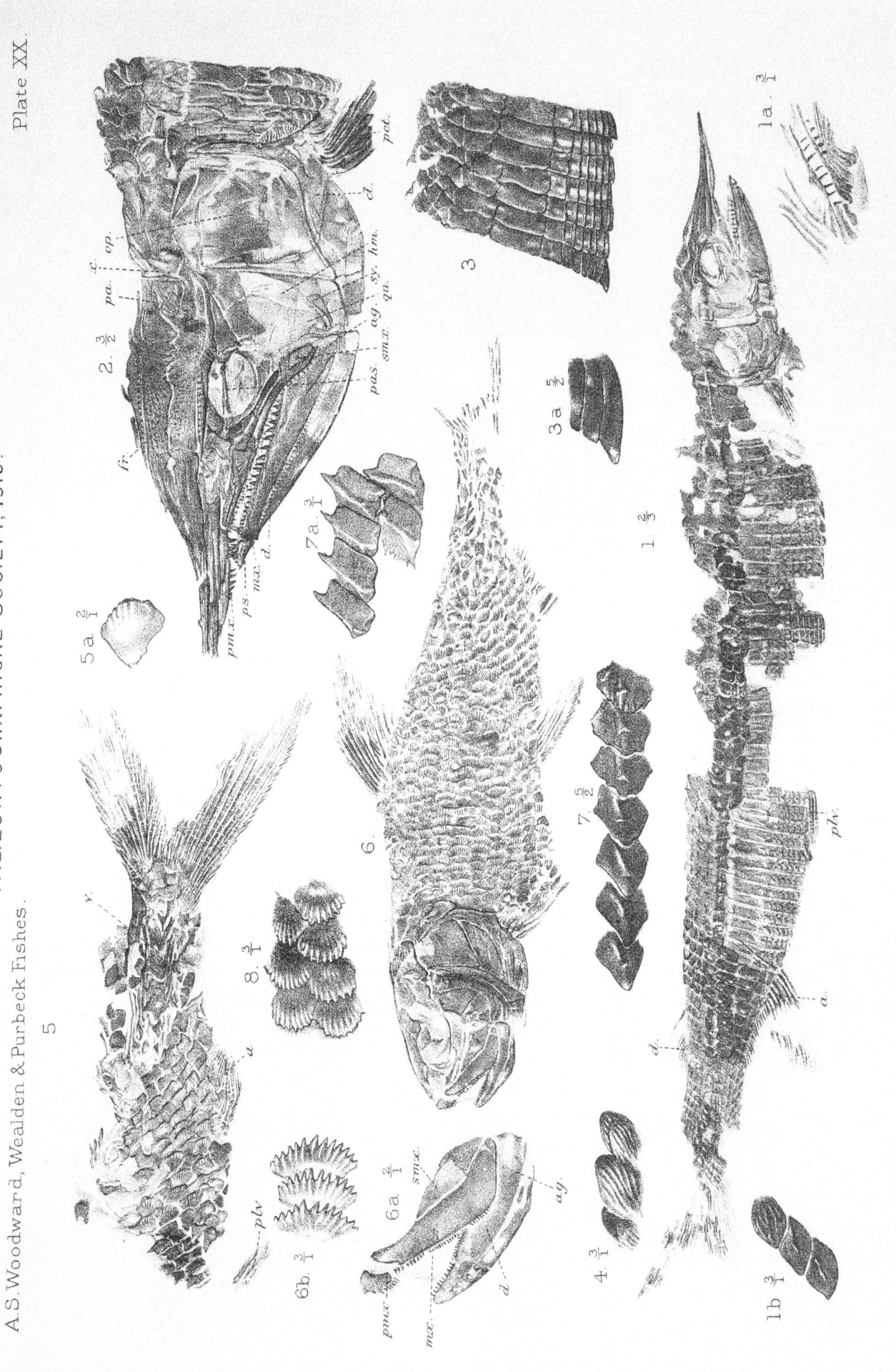

A. S. Woodward, Wealden & Purbeck Fishes.

G. M. Woodward, del et lith. Huth, imp.

1—4. Aspidorhynchus.   5—8. Pholidophorus

PLATE XXI.

FIG.   PAGE

1. *Belonostomus hooleyi*, sp. nov.; left flank-scale, outer view, lacking lower end and part of outer face, one-and-a-half times nat. size.—Wealden; Atherfield, Isle of Wight. The type specimen. Collection of Reginald W. Hooley, Esq., F.G.S.   100.

2. Ditto; dorsal scale, outer view, three times nat. size.—Wealden; Isle of Wight. Mantell Collection (B. M. no. 28419).   100.

3. Ditto (?); right flank-scale, outer view, one-and-a-half times nat. size.—Wealden; Sevenoaks, Kent. Sedgwick Museum, Cambridge.   101.

4. *Pholidophorus ornatus*, Agassiz; imperfect head and abdominal region, right lateral and ventral view, one-and-a-half times nat. size.—Middle Purbeck Beds; Swanage, Dorset. Museum of Practical Geology, London, no. 28439. *ag.*, angular; *br.*, branchiostegal rays; *cl.*, clavicle; *d.*, dentary; *iop.*, interoperculum; *mx.*, maxilla; *op.*, operculum; *orb.;* orbit; *pct.*, pectoral fin; *plv.*, pelvic fin; *pop.*, preoperculum; *scl.*, supraclavicle; *so.*, postorbitals; *sop.*, suboperculum.   103.

5. *Pholidophorus granulatus*, Egerton; greater part of fish, right lateral and ventral view, two-thirds nat. size, with some anterior flank-scales (5 a) and some caudal flank-scales (5 b), outer view, enlarged one-and-a-half times.—Ibid. Beckles Collection (B. M. no. P. 6379). *op.*, left operculum, inner view; *sop.*, left suboperculum, inner view.   106.

6. Ditto; portion of roof of skull, upper view, showing frontals (*fr.*), left parietal (*pa.*), and right squamosal (*sq.*), nat. size, with some anterior flank-scales (6 a) and caudal scales (6 b), inner view, enlarged one-and-a-half times.—Ibid. Enniskillen Collection (B. M. no. P. 3605).   106.

PALÆONTOGRAPHICAL SOCIETY, 1917.

A.S.Woodward, Wealden & Purbeck Fishes.

Plate XXI.

G.M.Woodward del et lith.

Huth,imp.

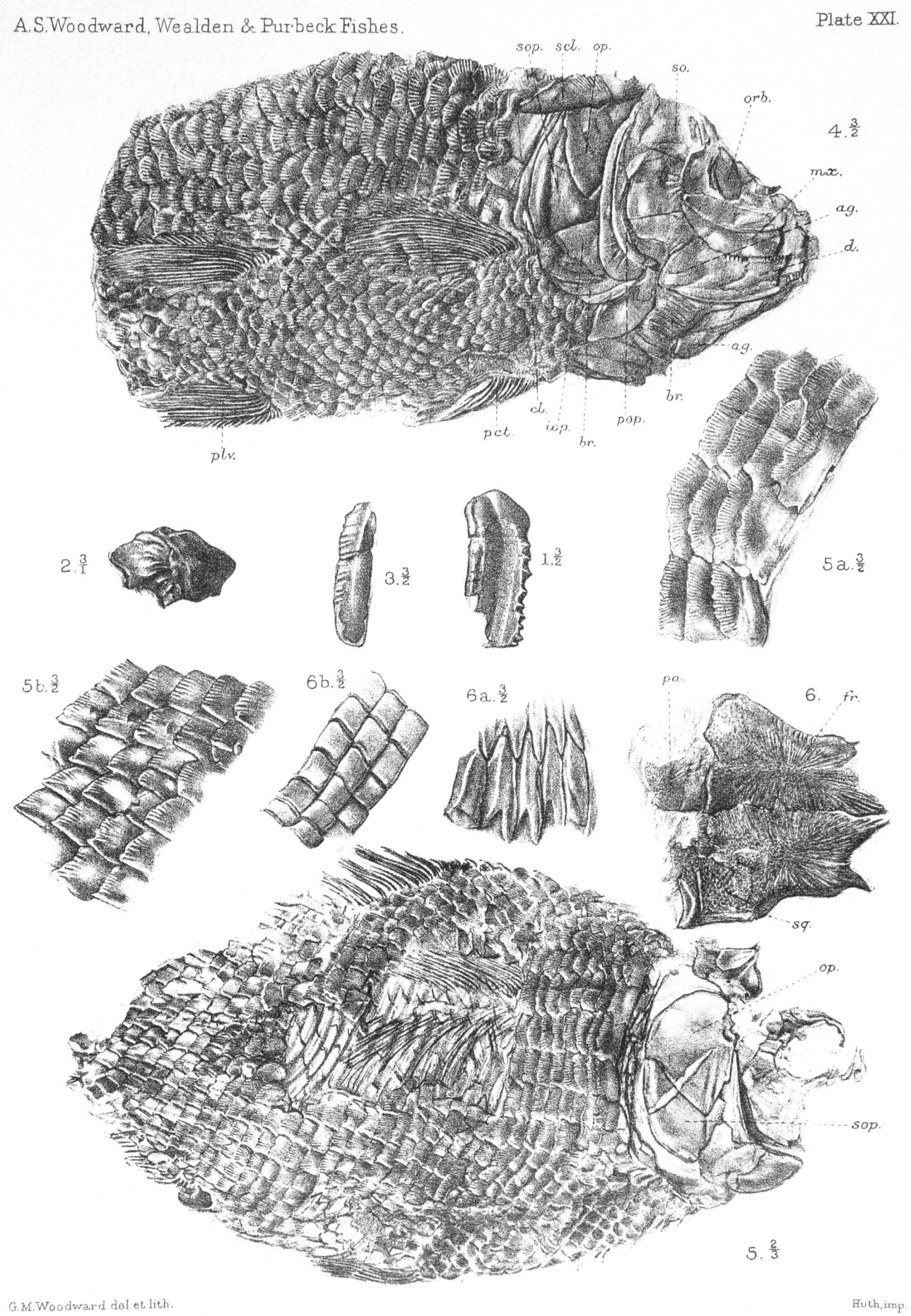

1-3. Belonostomus.   4-6. Pholidophorus.

PLATE XXII.

FIG.   PAGE.

1. *Pholidophorus purbeckensis*, Davies; imperfect fish, right lateral view, one-and-a-half times nat. size, with some anterior flank-scales (1 a) and caudal scales (1 b), outer view, enlarged three times.—Lower Purbeck Beds; Isle of Portland. The type specimen. Damon Collection (B. M. no. P. 6171). *pop.*, preoperculum; *scl.*, supraclavicle.   108.

2. Ditto; nearly complete fish, right lateral view, one-and-a-half times nat. size, with foremost anal fin-ray and fulcra (2 a) enlarged four times. —Ibid. British Museum, no. P. 8378.   108.

3. Ditto; small fish, right lateral view, twice nat. size.—Lower Purbeck Beds; Teffont, Vale of Wardour, Wiltshire. *iop.*, interoperculum; *op.*, upper part of operculum; *sop.*, suboperculum. P. B. Brodie Collection (B. M. no. P. 7640).   108.

4. *Pholidophorus brevis*, Davies; imperfect fish, left lateral view, one-and-a-half times nat. size.—Upper Purbeck Beds; Upway, near Weymouth. The type specimen. Egerton Collection (B. M. no. P. 1074).   110.

5. Ditto; head and anterior abdominal region, much broken by crushing, one-and-a-half times nat. size, with some flank-scales (5 a), inner view, enlarged twice.—Ibid. Enniskillen Collection (B. M. no. P. 3607).   110.

6. *Pholidophorus limbatus*, Agassiz; front of anal fin with fulcra, four times nat. size, for comparison with fig. 2 a.—Lower Lias; Lyme Regis, Dorset. Egerton Collection (B. M. no. P. 1047 b).

7. *Ceramurus macrocephalus*, Egerton; distorted fish, left lateral view, with counterpart of head (7 a), two-and-a-half times nat. size, and three caudal ridge-scales (7 b), enlarged ten times.—Middle Purbeck Beds; Dinton, Vale of Wardour, Wiltshire. The type specimen. Brodie Collection. (B. M. no. P. 7639). *cl.*, clavicle; *fr.*, frontal; *pa.*, parietal; *qu.*, quadrate; *scl.*, supraclavicle.   111.

8. *Pleuropholis*, sp.; immature fish, right lateral view, three times nat. size, with dorsal fin (8 a) enlarged five times.—Lower Purbeck Beds; Teffont, Vale of Wardour, Wiltshire. Rev. W. R. Andrews' Collection (B. M. no. P. 9846).   117.

PALÆONTOGRAPHICAL SOCIETY, 1917.

A.S.Woodward, Wealden & Purbeck Fishes.

Plate XXII.

G.M.Woodward, del.et lith.

Huth,imp.

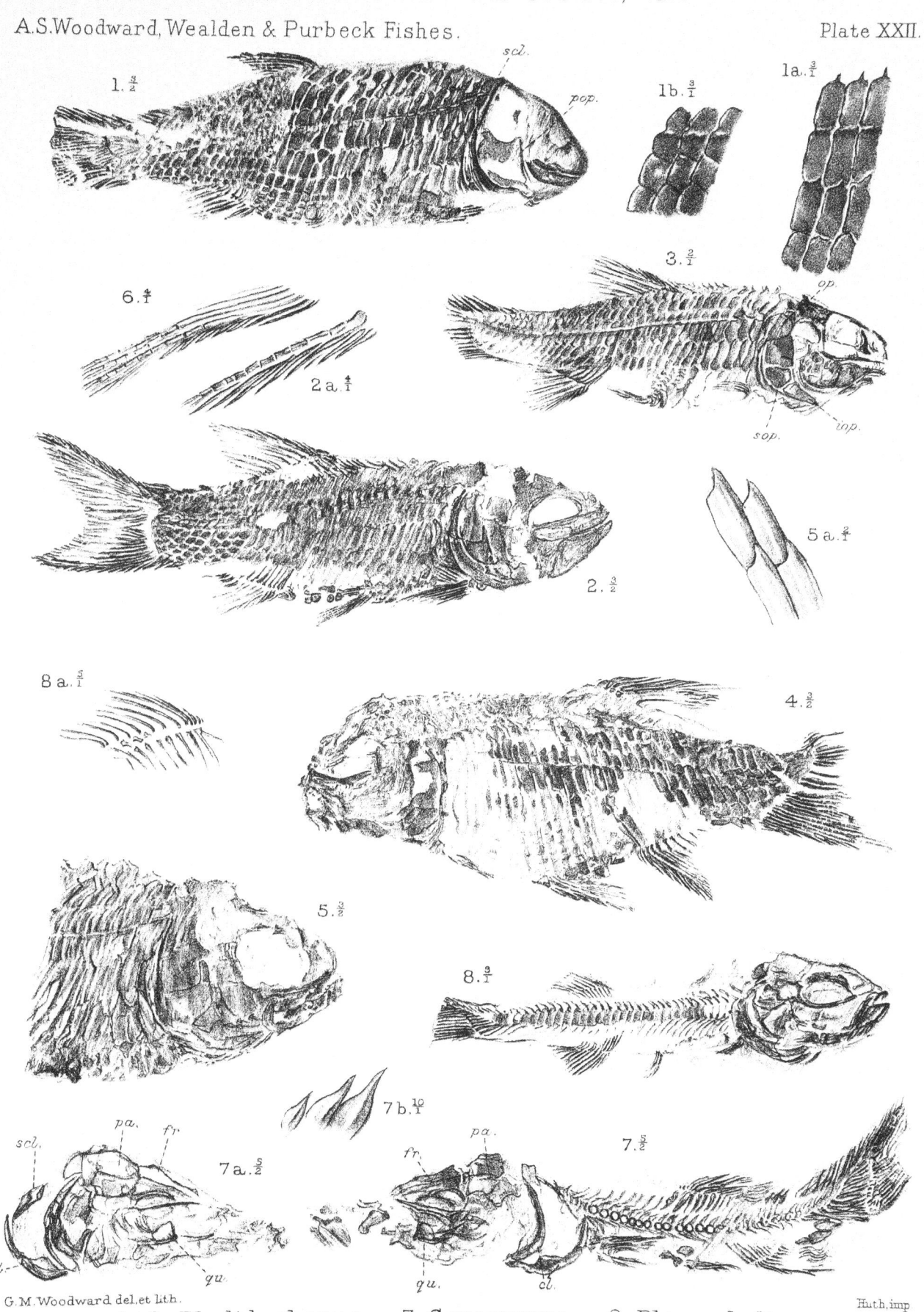

1–6. Pholidophorus.   7. Ceramurus.   8. Pleuropholis.

PLATE XXIII.

FIG.		PAGE.
1.	*Leptolepis brodiei*, Agassiz; very small immature fish, right lateral view, four times nat. size.—Lower Purbeck Beds; Vale of Wardour, Wiltshire. The type specimen of *Oxygonius tenuis*, Agassiz. P. B. Brodie Collection (B. M. no. P. 4730).	125.
2.	Ditto; immature fish shortened by distortion, right lateral view, five times nat. size.—Ibid. The type specimen of *Leptolepis nanus*, Egerton. P. B. Brodie Collection (B. M. no. P. 7637).	125.
3.	Ditto; fish lacking most of caudal fin, left lateral view, twice nat. size.—Ibid. The type specimen. P. B. Brodie Collection (B. M. no. P. 7635).	123.
4.	Ditto; distorted fish, three times nat size.—Ibid. P. B. Brodie Collection (B. M. no. P. 7635 a).	123.
5, 6.	Ditto; two well-preserved specimens of trunk, left lateral view, twice nat. size.—Lower Purbeck Beds; Lime Kiln Quarry, Teffont, Vale of Wardour, Wiltshire. Rev. W. R. Andrews' Collection (B. M. nos. P. 9847, 48). *c.*, coprolitic matter in intestine.	123.
7.	*Leptolepis*, sp.; right dentary bone, inner view, twice nat. size.—Wealden (Wadhurst Clay); Broad Oaks, Brede, near Hastings. Teilhard and Pelletier Collection (B. M. no. P. 11904).	125.
8.	*Pleuropholis formosa*, sp. nov.; nearly complete fish, left lateral view, twice nat. size.—Lower Purbeck Beds; Teffont, Wiltshire. The type specimen. British Museum, no. P. 10986.	115.
9.	Ditto; head, left lateral view, three times nat. size.—Ibid. British Museum, no. P. 10955. *ecpt.*, ectopterygoid; *enpt.*, entopterygoid; *md.*, mandible; *op.*, operculum; *pas.*, parasphenoid; *pop.*, pre-operculum.	116.
10.	Ditto; foremost pectoral fin-ray bearing expanded fulcra, enlarged ten times.—Ibid. British Museum, no. P. 9851.	116.
11.	Ditto; part of anal fin, showing supports, five times nat. size.—Ibid. British Museum, no. P. 10986.	116.
12.	*Pleuropholis crassicauda*, Egerton; imperfect fish, right lateral view, twice nat. size.—Middle Purbeck Beds; Durdlestone Bay, Swanage, Dorset. The type specimen. P. B. Brodie Collection (B. M. no. P. 7647).	118.
13.	Ditto; fish with incomplete fins, left lateral view, twice nat. size.—Ibid. British Museum, no. 43615.	118.

PALÆONTOGRAPHICAL SOCIETY, 1917.

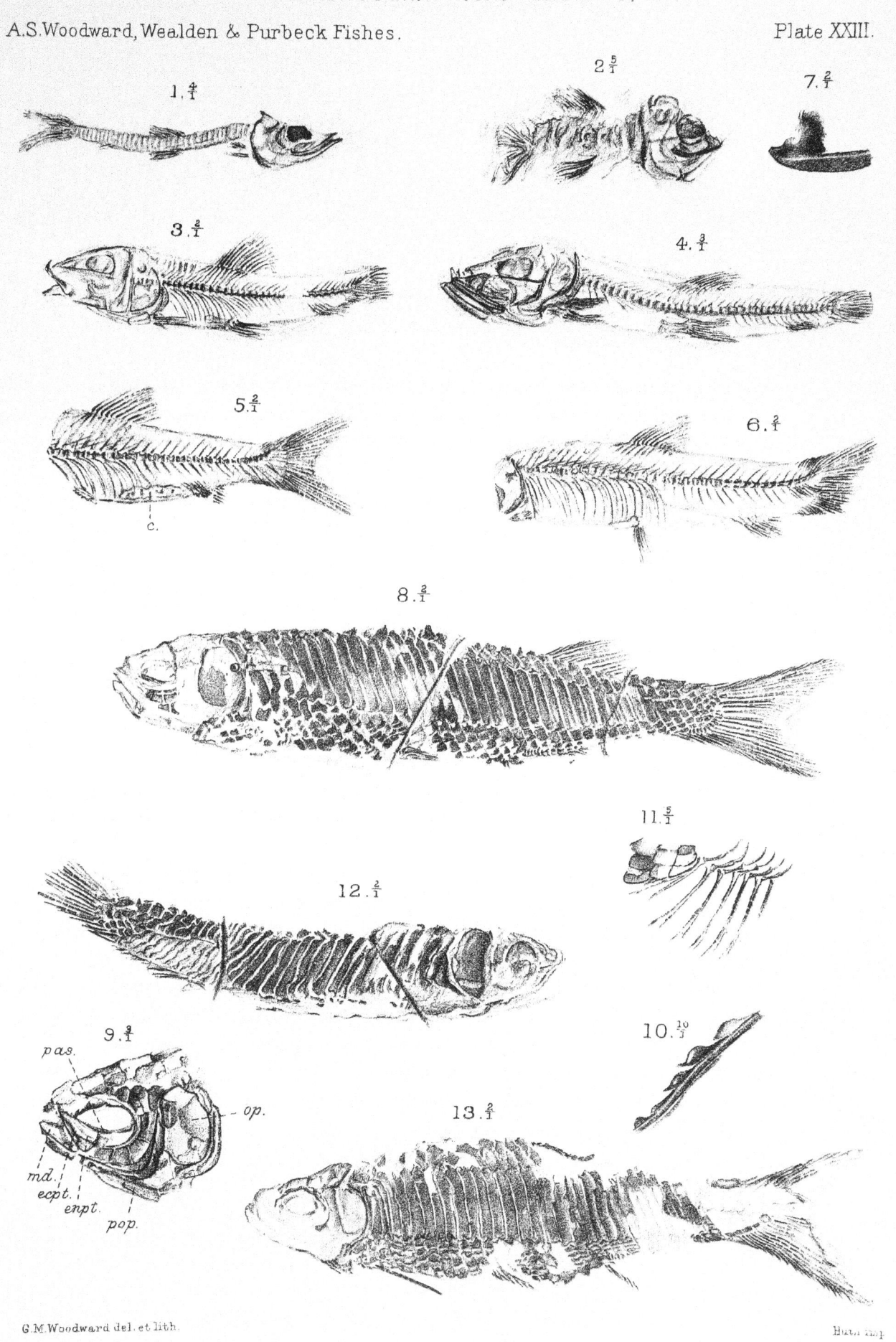

A.S.Woodward, Wealden & Purbeck Fishes. Plate XXIII.

G.M.Woodward del. et lith.

1–7. Leptolepis.   8–13. Pleuropholis.

PLATE XXIV.

FIG.   PAGE.

1. *Pleuropholis longicauda*, Egerton; head and abdominal region, left lateral view, nat. size, with some dorsal scales (1 *a*) and some ventral scales (1 *b*) enlarged four times.—Middle Purbeck Beds; Swanage, Dorset. British Museum, no. 40724.   120.

2. Ditto; imperfect fish, left lateral view, nat. size, with some scales (2 *a*) enlarged twice.—Ibid. British Museum, no. P. 7664.   120.

3. *Pachythrissops lævis*, sp. nov.; roof of skull, lacking end of snout, nat. size.—Ibid. British Museum, no. P. 12212. *epo.*, epiotic; *fr.*, frontal; *pa.*, parietal; *ptf.*, postfrontal (sphenotic); *socc.*, supra-occipital; *sq.*, squamosal.   129.

4. Ditto; remains of head and three anterior vertebræ, right lateral view, two-thirds nat. size, with some lower teeth (4 *a*) enlarged twice.—Ibid. Portion of specimen figured as *Lepidotus minor* by L. Agassiz, Poiss. Foss., vol. ii, pt. i (1844), pl. xxix *c*, fig. 12. Enniskillen Collection (B. M. no. P. 4219). *bh.*, basihyal; *br.*, branchiostegal rays; *ch.*, ceratohyal; *d.*, dentary; *ecpt.*, ectopterygoid; *eph.*, front of epihyal; *g.r.*, gill-rakers on gill-arch.   129.

5. Ditto; middle portion of trunk, right lateral view, nat. size.—Ibid. British Museum, no. 44845. *d.*, dorsal fin-supports and fin; *plv.*, pelvic fin-supports with fins.   132.

6. *Pachythrissops vectensis*, A. S. Woodward; portion of head, left lateral view, with lower expansion of right preoperculum (6 *a*), one-quarter nat. size.—Weald Clay; Atherfield, Isle of Wight. Collection of Reginald W. Hooley, Esq., F.G.S. *op.*, operculum; *pop.*, pre-operculum; *so.*, postorbital cheek-plates.   134.

7. Ditto; scale, nat. size.—Ibid. Collection of Reginald W. Hooley, Esq., F.G.S.   136.

PALÆONTOGRAPHICAL SOCIETY, 1917.

A.S. Woodward, Wealden & Purbeck Fishes. Plate XXIV.

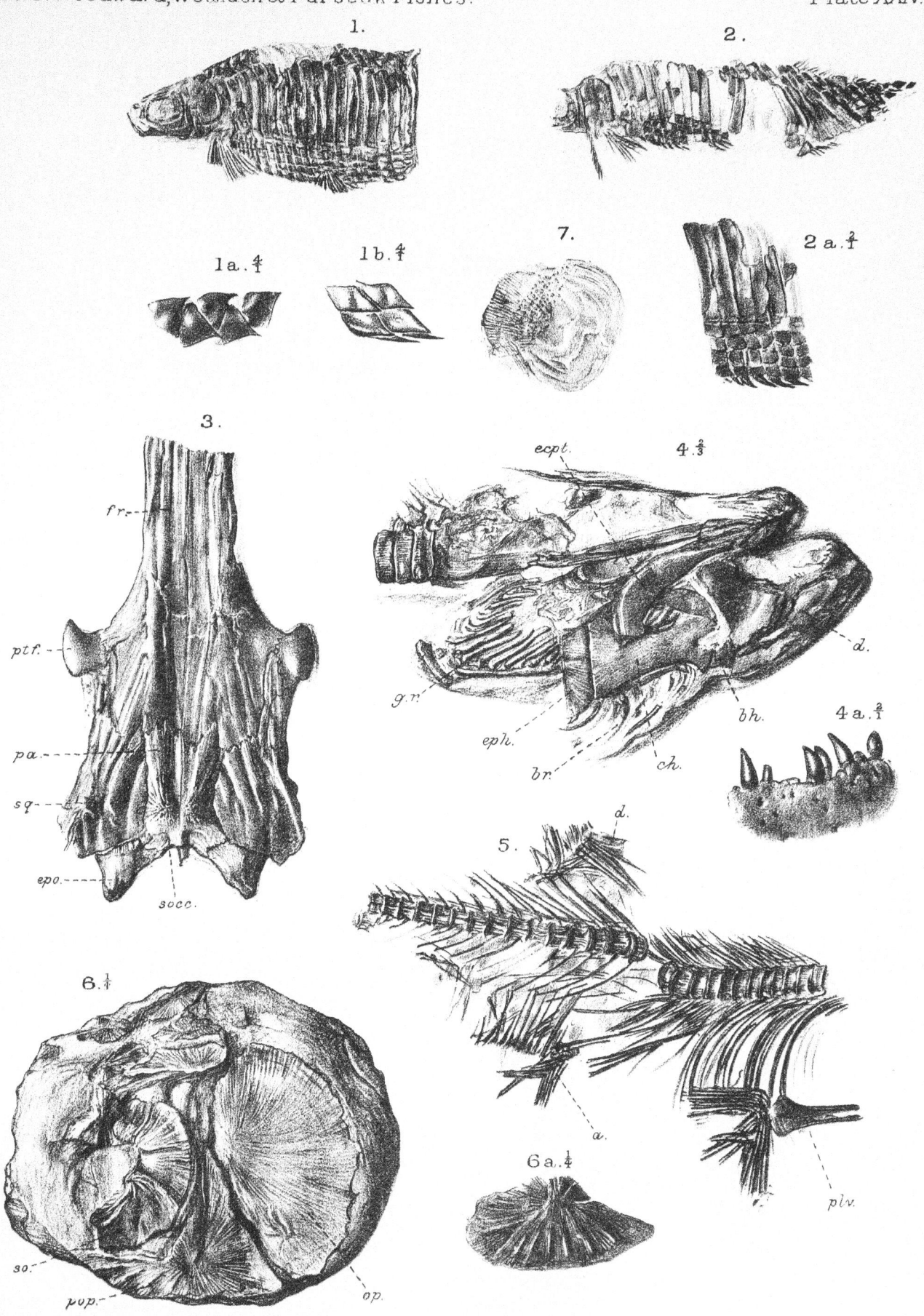

G.M. Woodward del. et lith. Huth imp.

1, 2. Pleuropholis.   3–7. Pachythrissops.

PLATE XXV.

FIG.   PAGE.

1. *Pachythrissops lævis*, sp. nov.; nearly complete fish, left lateral view, nat. size, with part of operculum (1 a) enlarged six times to show slime-pit.—Middle Purbeck Beds; Swanage, Dorset. The type specimen. British Museum, no. 40433. *br.*, branchiostegal rays; *cl.*, clavicle; *op.*, operculum; *pop.*, preoperculum; *scl.*, supraclavicle; *sop.*, suboperculum.   129.

2. Ditto; left dentary, outer view, one-third nat. size.—Ibid. British Museum, no. 36083.   131.

3. Ditto; vertebral centrum, end view to show perforation by notochord, nat. size.—Ibid. British Museum, no. 21974.   132.

4. *Pachythrissops vectensis*, A. S. Woodward; imperfect fish, right lateral view, one-quarter nat. size.—Weald Clay; Atherfield, Isle of Wight. Collection of Reginald W. Hooley, Esq., F.G.S. *a.*, portion of anal fin; *ag.*, angular; *br.*, branchiostegal rays; *cl.*, clavicle; *ecpt.*, ectopterygoid; *mx.*, maxilla; *pct.*, pectoral fins; *plv.*, pelvic fins; *qu.*, quadrate; *scl.*, supraclavicle; *sy.*, symplectic; *x.*, probably posttemporal.   133.

5. Ditto; upper part of left hyomandibular, outer view, one-half nat. size.—Wealden; Isle of Wight. British Museum, no. 42013. *p.*, process for support of operculum.   134.

PALÆONTOGRAPHICAL SOCIETY, 1917.

Plate XXV.

A. S. Woodward, Wealden & Purbeck Fishes

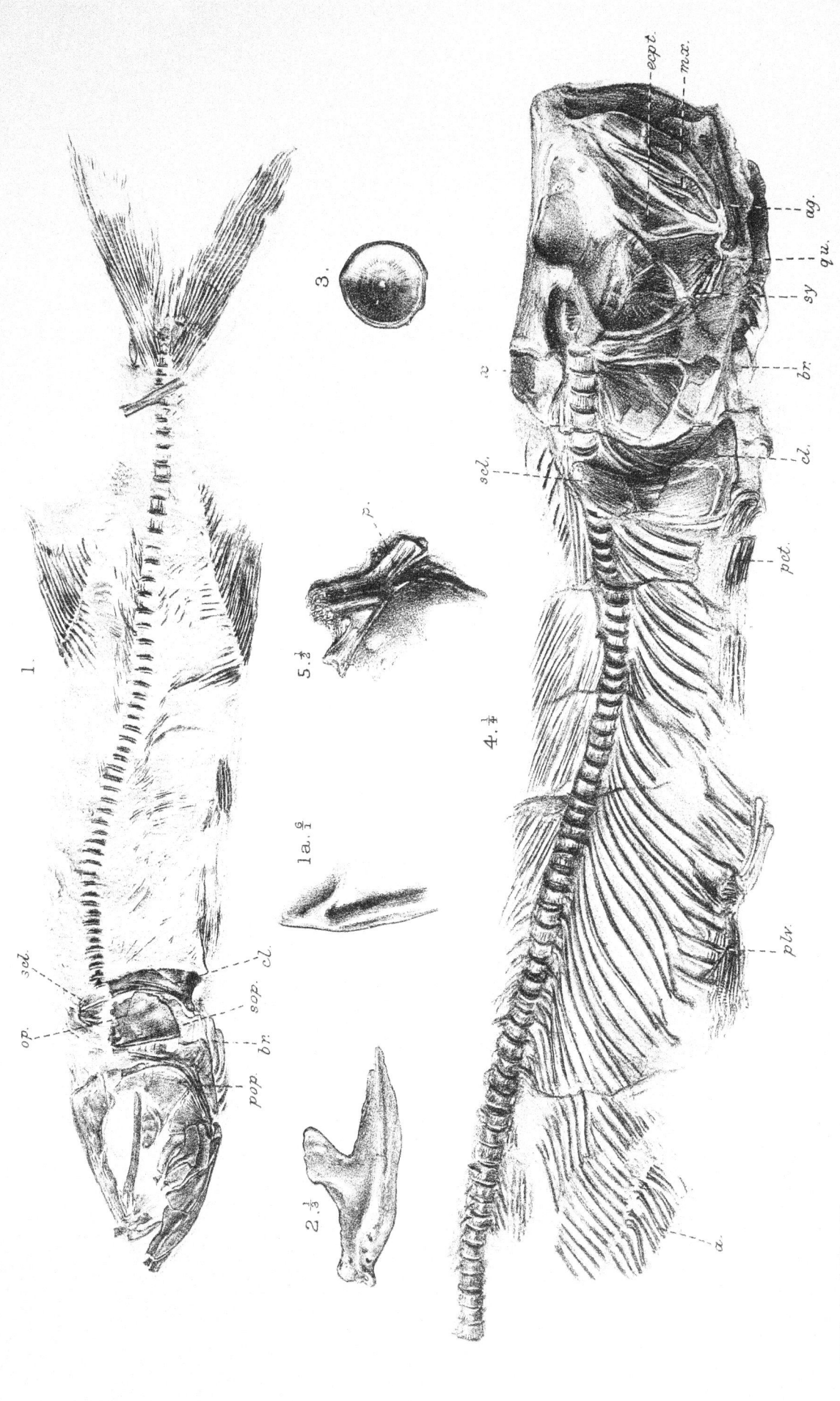

Pachythrissops.

G. M. Woodward del. et lith.   Huth imp.

PLATE XXVI.

FIG.   PAGE.

1. *Thrissops curtus*, sp. nov.; right lateral view of fish, nat. size.—Lower Purbeck Beds; Isle of Portland. The type specimen. British Museum, no. P. 10612.   137.

2. *Thrissops molossus*, sp. nov.; left lateral view of fish, two-fifths nat. size.—Middle Purbeck Beds; Swanage. British Museum, no. P. 417 *a*.   138.

3. *Hybodus basanus*, Egerton; cranium, upper and (3 *a*) posterior views, one-half nat. size.—Weald Clay; Atherfield, Isle of Wight. Collection of Reginald W. Hooley, Esq., F.G.S. *f.m.*, foramen magnum; *n.*, pit for notochord; *p.f.*, posterior fossa or fontanelle; *v.*, pair of pits into which the vagus nerves probably opened.   139.

4. *Lepidotus minor*, Agassiz; imperfect fish, left lateral view, showing dorsal fin, one-third nat. size.—Middle Purbeck Beds; Swanage. British Museum, no. P. 12211.   140.

PALÆONTOGRAPHICAL SOCIETY, 1917.

A. S. Woodward, Wealden & Purbeck Fishes.

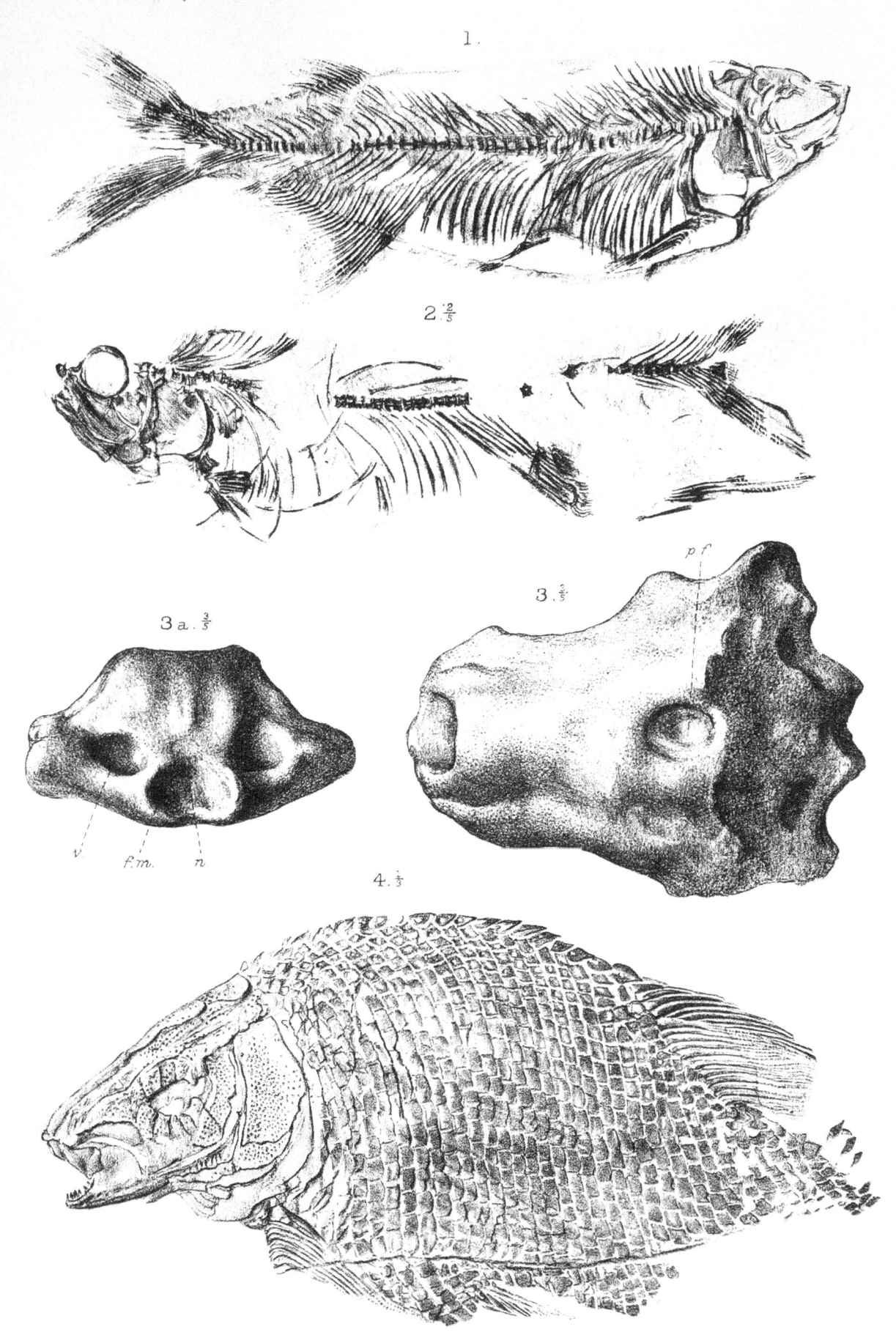

Plate XXVI.

G. M. Woodward del. et lith.

Huth imp.

1, 2. Thrissops. 3. Hybodus 4. Lepidotus.

Palæontographical Society, 1915.

THE
FOSSIL FISHES

OF THE

ENGLISH

WEALDEN AND PURBECK FORMATIONS.

BY

ARTHUR SMITH WOODWARD, LL.D., F.R.S.,

KEEPER OF THE DEPARTMENT OF GEOLOGY IN THE BRITISH MUSEUM; SECRETARY OF THE
PALÆONTOGRAPHICAL SOCIETY.

PART I.

PAGES 1—48, PLATES I—X.

LONDON:
PRINTED FOR THE PALÆONTOGRAPHICAL SOCIETY.
OCTOBER, 1916.

PRINTED BY ADLARD AND SON AND WEST NEWMAN, LONDON AND DORKING.

Palæontographical Society, 1916.

THE FOSSIL FISHES

OF THE

ENGLISH

WEALDEN AND PURBECK FORMATIONS.

BY

ARTHUR SMITH WOODWARD, LL.D., F.R.S.,

KEEPER OF THE DEPARTMENT OF GEOLOGY IN THE BRITISH MUSEUM; SECRETARY OF THE
PALÆONTOGRAPHICAL SOCIETY.

PART II.

PAGES 49—104, PLATES XI—XX.

LONDON:
PRINTED FOR THE PALÆONTOGRAPHICAL SOCIETY.
FEBRUARY, 1918.

PRINTED BY ADLARD AND SON AND WEST NEWMAN, LTD., LONDON AND DORKING.

Palæontographical Society, 1917.

THE
FOSSIL FISHES

OF THE

ENGLISH

WEALDEN AND PURBECK FORMATIONS.

BY

ARTHUR SMITH WOODWARD, LL.D., F.R.S.,

KEEPER OF THE DEPARTMENT OF GEOLOGY IN THE BRITISH MUSEUM; SECRETARY OF THE
PALÆONTOGRAPHICAL SOCIETY.

PART III.

Pages 105—148, i—viii, Plates XXI—XXVI (including Title-page and Index).

LONDON:
PRINTED FOR THE PALÆONTOGRAPHICAL SOCIETY.
April, 1919.

PRINTED BY ADLARD AND SON AND WEST NEWMAN, LTD., LONDON AND DORKING.